PLANNING IN BRITAIN

PLANNING IN BRITAIN

Understanding and Evaluating
the Post-war System

Andrew W. Gilg

SAGE Publications
London ● Thousand Oaks ● New Delhi

This book is dedicated to the Scottish Rugby Union teams who beat
England 19 points to 13 in April 2000 and South Africa 21 points to 6 in
November 2002 and who gave me the inspiration to write this book after
two setbacks to my self-confidence

First published 2005

SAGE Publications Ltd
1 Oliver's Yard
55 City Road
London EC1Y 1SP

SAGE Publications Inc.
2455 Teller Road
Thousand Oaks, California 91320

SAGE Publications India Pvt Ltd
B-42, Panchsheel Enclave
Post Box 4109
New Delhi 110 017

British Library Cataloguing in Publication data

A catalogue record for this book is available from the British Library

ISBN 0 7619 4929 1
ISBN 0 7619 4930 5 (pbk)

Library of Congress Control Number available

Typeset by M Rules
Printed and bound in Great Britain by Athenaeum Press, Gateshead

Contents

List of boxes   vii
List of figures   viii
List of tables   x
Acknowledgements   xi
Introduction   1

1. How and Why the Planning System Evolved   5

Historical evolution of plan making   5
The development control system   32

2. Issues in Evaluation   49

Public policy within the context of concepts of power and society   50
Why society may want planning to intervene, with what objectives
and by how much?   55
Review of alternative delivery systems for planning   62
Problems involved with evaluating land use planning   67
Review and choice of evaluation methods   73

3. Evaluating Planning Outputs   85

Plan making   85
Development control   96

4. Evaluating Planning Outcomes and Impacts   111

Managing land use change outcomes   112
Selected studies of outcomes and impacts   130

5. A Set of Evaluations and the Way Forward   153

Reminder of the issues and problems in evaluating planning   153
Four types of evaluation   154

A hybrid evaluation   173
The future as a mirror of the past   185
Personal postscript   195

Bibliography   199
Index   207

List of boxes

1.1	Key organisations involved in creating the planning system	7
1.2	Summary of three key reports in the 1960s	12
1.3	Split between planning functions from 1974 onwards	12
1.4	Functions of Structure and Local Plans as set out in *Manual on Development Plans: Form and Content* (1970)	15
1.5	Policy guidance to local planning authorities	23
1.6	The scope of Development Plans in England	26
1.7	Development Plans in 2002	31
1.8	Example of advice on development control	35
1.9	Key stages in the development control process	38
1.10	Case study of an appeal decision: golf course	41
1.11	Case study of a planning appeal: oil refinery	42
2.1	Political history of post-war Britain	53
2.2	Government objectives for planning in 2002	58
2.3	Benchmarking planning	66
2.4(a)	Hybrid (egg-timer) approach to evaluation	80
2.4(b)	Hybrid (egg-timer) approach to evaluation	81
3.1	Development Plans: common criticisms and reform potential	95
4.1	An editorial in the *Sunday Times*, 30 June 2002	123
4.2	Policy changes announced in July 2002	124
5.1	Mock exam questions typifying changing attitudes to planning	159
5.2	Differences between conflict management and place making	171
5.3	Best Value performance indicators for planning	185
5.4	Reflections on planning on a weekend away in July 2004	196

List of figures

1.1 New administrative areas as at 1 April 1974 (Gilg, 1978) 13
1.2 The process of Development Plan preparation in the 1980s and 1990s (Thomas, 1996) 16
1.3 Typical arrangements at an Examination in Public (adapted from Murdoch and Abram, 2002) 17
1.4 Changes in housing provision between Structure Plans (adapted from Coates, 1992) 18
1.5 Negotiations in the Development Plan process (Claydon, 1996) 19
1.6 Strategic planning frameworks in England, 1980–95 (adapted from Thomas and Roberts, 2000) 22
1.7 Planning authorities in the United Kingdom (adapted from Cullingworth and Nadin, 2002) 25
1.8 The hierarchy of planning policy guidance in Wales in the mid-1990s (Jarvis, 1996) 30
1.9 Negotiations in the development control process (Claydon, 1996) 34
2.1 Flow chart from objectives to evaluation 50
2.2 The triangle of conflicting goals for planning (adapted from Campbell, 1996) 56
2.3 An ideal planning system? (Gilg, 1996) 63
2.4 Unintended policy outcomes (Schrader, 1994) 68
2.5 The planning constraints coffin (Gilg, 1996) 69
2.6 The planning process and those who may influence it (Lichfield, 1996) 71
2.7 An adapted Toulmin–Dunn model for evaluating public policy (adapted from Gasper and George, 1998) 75
2.8 A flow chart for evaluating the planning system (adapted from Department of the Environment, 1992) 77
2.9 A framework for developing indicators for land use planning (adapted from Morrison and Pearce, 2000) 77
2.10 A framework for evaluating the planning system and its individual components (author) 78
2.11 The hybrid approach to evaluating land use planning (author) 82
3.1 Agent interactions in the planning process (Short et al., 1987) 98
3.2 Development Plan policies in Buckinghamshire (Murdoch and Marsden, 1994) 100
3.3 Appeal rate, by type of local authority (adapted from Wood, 2000) 104
3.4 Appeal pressure and support at appeal, by district (adapted from Wood, 2000) 105
3.5 The cognitive continuum of decision making (adapted from Willis and Powe, 1995) 107

4.1   Regional conversions of farmland to urban use, England and Wales, 1945–65 (Best and Champion, 1970)   113

4.2   Total houses built in Britain (DETR statistics), 1920–2001 (adapted from *The Times*, 3 May 2002)   115

4.3   Megalopolis: land use, showing degree of urbanisation *c.* 1960 (Hall, 1974)   118

4.4   Population change, 1991–2001 (adapted from Office of National Statistics map   128

4.5   Conceptualisation of population and settlement change around major cities (Gilg, 1996 from Moss, 1978)   129

4.6   Wheel method of analysing planning applications (Murdoch and Marsden, 1994)   132

4.7   The relationship of decisions to Local Plans (adapted from Pountney and Kingsbury, 1983)   133

4.8   Planning applications per thousand population (Curry and McNab, 1986)   135

4.9   Approval and refusal ratios in North Yorkshire (Midgeley, 2000)   136

4.10   Key settlement policy for Devon (Gilg, 1978)   138

4.11   Planning constraints in east Devon (Gilg and Blacksell, 1981)   139

4.12   Example of an estate of modern houses tacked on to a village (Gilg and Blacksell, 1981)   140

4.13   Location of permitted and refused planning applications in East Sussex, 1975–76 (adapted from Anderson, 1981)   143

4.14   Stocks of planning permission, 1981–91 (adapted from Monk and Whitehead, 1999)   147

4.15(a) Housing land and (b) house prices, 1981–91 (adapted from Monk and Whitehead, 1999)   147

4.16   New housing completions in (a) north Hertfordshire, (b) south Cambridgeshire and (c) Fenland, 1981–91 (adapted from Monk and Whitehead, 1999)   149

4.17   Settlement policies in Structure Plans (adapted from Cloke and Shaw, 1983)   151

5.1   Evolution of design styles in post-war Britain (*Planning*, 409 13 March 1981)   158

5.2   Relationship between theoretical approaches to planning (Rydin, 1993)   165

5.3   Theoretical approaches based on a view of the economic process (Rydin, 1993) 166

5.4   The development of planning policy and theory (Rydin, 1998)   167

5.5   The evolution of regulatory purposes in the twentieth century (adapted from Healey, 1998)   170

5.6   Conflict management and place making as alternative strategies for planning (adapted from Healey, 1998)   172

5.7   The projected rate of change to urban uses, 1991–2016 (adapted from Bibby and Shepherd, 1997)   186

5.8   The projected rate of urban growth, 1991–2016 (adapted from Bibby and Shepherd, 1997)   187

5.9   Revised plan-making system envisaged by the 2001 Green Paper (author)   193

List of tables

1.1	Examples of changes made by the Secretary of State of the Environment to structure plans after the EIP	17
1.2	Stages and timing in preparation of the first generation of Local Plans	21
1.3	Examples of Local Plan preparation and modifications	21
1.4	Structure of local government in 2002	27
1.5	Main concerns and objectives of RPG in England and Wales	28
1.6	Key regional planning documents in England (outside London), 2002	29
1.7	Conditions on permissions used in rural East Sussex	38
1.8	Reasons for refusal of planning permission used in rural East Sussex	40
2.1	UK headline indicators for sustainable development	74
3.1	Landowner involvement in plan making	91
4.1	Land use changes in England and Wales, 1933–1963	112
4.2	Annual average net losses of farmland to urban use, England and Wales	113
4.3	Previous use of land changing to residential use, England, 1985–95	114
4.4	Centralisation and decentralisation in the twentieth century	117
4.5	Percentages of land forecast in 1985 to be surplus to agriculture by 2000	119
4.6	Disputed housing forecasts in the south-east of England	123
4.7	Components of population change, 1981–91, by region and change, 1981–2001	126
4.8	Population change by top and bottom areas, 1981–91	127
4.9	The relationship between data sources and analytical methodologies	131
4.10	Development pressures in England and Wales	134
4.11	The relationship between planning applications and Development Plan policies in south-east Devon	137
4.12	Decisions on development control applications in four areas in Devon	140
4.13	Land use changes and their relation to land use planning policies	141
4.14	Planning applications, by area, in East Sussex	142
4.15	Representations on planning application to convert a cinema to a retirement residence	144
4.16	House prices related to average wages, by region	146
5.1	Changing attitudes towards planning under the New Right	161
5.2	New Right and New Labour comparisons	161
5.3	The changing orthodoxy of planning	162
5.4	A typology of planning problems, policy goals and policy outcomes	163
5.5	A typology of planning styles	164
5.6	Characteristics of four planning styles identified by Brindley *et al.*	164
5.7	Environment or environment-related plans in England	190

Acknowledgements

I am grateful to all the staff at Sage who have supported me through this project, notably Robert Rojek, Vanessa Harwood, Fabienne Pedroletti and David Mainwaring. I am also indebted to the three initial referees who made very useful comments on the first draft especially with regard to dividing the evaluation between procedures and impacts. I am also grateful to thirty generations of students who have sat through the lectures on which this text is based and unwittingly aided its improvement by their essay and examination answers. A big thank-you to the two cartographers who modified and made big improvements to many of the figures, Sue Rouillard and Helen Jones. Thanks are also due to my colleagues on the Policy Council of the Town and Country Planning Association, who represent a great cross-section of planning skills and cognate professions.

I am also grateful to an anonymous design editor who made many thoughtful suggestions for improving the book at the second draft stage. Finally I am very grateful to Mike Kelly, a long-term collaborator, for reading the first draft of the book from the point of view of a practising planner. Mike is a senior planner at North Devon District Council and he has been influential in developing my knowledge of planning since 1990, both through our joint research but also through training courses we have given to local authority councillors.

While acknowledging the help of many people, any errors are of course my own and the opinions in this book are my own and do not reflect the views of any organisation.

This book was written at a time of several crises in my life and the chance to escape into the apparently more logical world of planning evaluation provided some much needed stability at each of the three draft stages (summer 2002, winter 2003 and spring–summer 2004). Two planned creations, the golf courses at Crediton in Devon and Crans-sur-Sierre in Switzerland, also provided moments of inspiration. But most of all my immediate family, Joyce, Julie and Alastair, continue to form the basis of my creativity.

A.W.G.
Exeter, July 2004

Introduction

It certainly was not the intentions of the founders of planning that people should live cramped lives in houses destined for premature slumdom far from urban services or jobs, or that city dwellers should live in blank cliffs of flats far from the ground without access to play space for their children. Somewhere along the way a great deal was lost, a system distorted and the great mass of people betrayed.

Hall (1974) p. 407

The planning system has organised the process of suburbanisation but not resisted it.

Healey (1983) p. 41

These two rather bleak quotations provide the sort of evaluation that this books seeks to develop further in the context of a British planning system that has managed land use change for nearly sixty years. Indeed, well over half the British-built environment has been constructed under its auspices. It is thus very surprising to find that there have been very few evaluations of the system, although there are a few histories of the system, notably the companion Sage book by Stephen Ward (2004), but this is limited to urban change. A new book by Carmona and Sieh (2004) due out one month after this book was completed, promised to cover one aspect of evaluation, the quality of the processes, but this would fall far short of the full and holistic evaluation attempted by this book. A rare example of such an attempt is provided by my book *Countryside Planning*, (1996), which examined the effect of a very wide range of policies on agriculture, forestry, settlement, recreation and conservation in rural areas.

This new book concentrates on the town and country planning system derived from a series of Town and Country Planning Acts between 1947 and 2004 and referred to henceforward as either land use planning or planning. It is also assumed from now onwards that planning refers to 'land use planning in Britain between 1947 and today' unless otherwise stated. As the two quotations at the outset make clear, the twin pillars of planning have been (1) to manage the process of urban development, notably the provision of housing, and (2) to protect the countryside and heritage areas from undesirable change.

Thus this book concentrates on how the planning system has been used to create the residential environment of urban and rural areas and to protect townscapes and the countryside. Other issues are considered where they have an impact on these issues or where planning decisions have had an impact elsewhere, for example house price inflation. But it is very important to remember that the Town and Country Planning Acts by and large restrict planning to *land use issues* and though too many ivory-tower academics would wish it otherwise in their theoretical accounts planning is rooted in the physical reality of bricks and mortar. Accordingly, this book focuses on a neglected aspect of planning, the pragmatic management of change from one land use to another, because that is the bedrock of the planning system. Nonetheless, this management takes place in the wider context of society, so the book addresses the following key generic issues:

1   To what extent is planning a mirror of British society or a catalyst for social change?
2   Is its key role thus to manage change with the minimum of conflict and uncertainty but with maximum efficacy as a decision-making process?
3   Are its dual objectives, to allow economic development, notably housing, without undue harm to the environment, realistically achievable without damage to one or the other?
4   To what extent is it acceptable that planning evaluates itself through its cyclical process of plan preparation or should it be more open to wider evaluations by other agencies?

And the following specific issues:

1   Are land use plans the most efficient way of promoting land use policies?
2   To what extent should development control decisions be based on land use policies?
3   To what extent can the outcomes and impacts of planning decisions be disentangled from other decisions in society in order to isolate the effect of planning on environmental and social change?

This book is thus about the tools that planners have been given, their efficacy or otherwise, how they have chosen to use them and, most important, what impacts there have been. It uses the three core issues of urban containment, settlement provision and landscape protection to evaluate how the land use transition to residential environments has occurred and what the wider impacts have been, notably in terms of traffic growth and house prices.

Such an evaluation does, however, have three big problems. First, is there a straight-line link between cause and effect? Second, the problem of by who and from what stance the evaluation is made, and third, the 'what if' or 'birth control' problem.

First, there is the assumption that there is a straight link between cause and effect. In this respect there is a clear link between evaluation and planning, since both are modernist concepts. Planning developed in common with modernism in the belief that the world could be made better by the application of knowledge through scientific principles and methods. At the most ambitious level it attempted to completely control things, for example by building New Towns. Less ambitiously, planning sold itself as a rational alternative to the failures of the free market to provide work, housing and a good environment. However, the real world is very complex and chaotic and attempts to understand, let alone control, it have been problematic, to say the least. In particular planners have neither the perfect knowledge required nor the perfect system to control the environment. Thus *planning can be considered to be the art of making sufficient decisions from insufficient information*. Accordingly, planners have pragmatically retreated into a mantra of managing change and resolving conflicts in a society dominated by capitalism and short-term greed captured in the aphorism 'God make me good, but not yet'.

In some senses this religious allusion is appropriate, since planners are trying to create a better world, albeit not the utopia or 'Garden of Eden' that they might once have done.

Planners attempt to improve things by two main mechanisms: (1) a vision of the future, usually in the form of a plan – hence 'planning'; (2) by attempting to manage change in accordance with the plan, mainly by the power to refuse or impose conditions on applications to development land, usually known as 'development control'. Accordingly, the book is structured in detail by these two different activities

Second, evaluation is often dependent on the position of the evaluator. Because planning makes decisions that affect people differently it will always create winners and losers, though planners are increasingly trying to create win–win solutions for society by using the principles of sustainable development. In particular, planners are attempting to manage change by reconciling the conflicts in the society–economy–environment triangle. This may not, however, be realistically achievable.

Third, and perhaps the biggest problem in evaluating planning, is the so-called 'birth control' or 'what-if' problem. In other words, planning has prevented many developments even being considered, let alone built, and therefore its impact can never be truly evaluated. There are thus a number of very different strategies that might have been followed and so planning might best be evaluated by considering what *actually* happened in the context of what *might* have happened if different interest groups had been in power.

More prosaically, planning exists within the social systems that gave it life and it can be evaluated in terms of its relation to power structures in society. Thus we can evaluate the way the system evolved in relation to social and economic change. Then we can evaluate planning as that subset of society that seeks to provide a service to manage environmental change at the generic level. Then we can evaluate the methods that society has given planning in terms of their effectiveness and efficiency as tools to do a job. Then we can evaluate the wider outcomes and impacts that have arisen from planning decisions. Finally, we can see if we have sufficiently robust and accurate evaluation systems to relate the decisions made by the planning decision to the way in which our environment has changed.

Accordingly the book is structured as chapters:

1   An evaluation of how the planning system evolved in terms of historical changes in society and the economy.
2   Issues in evaluating the system relating to planning's relationship to society and power structures, alternative planning procedures and problems in evaluation, which collectively demand a hybrid approach.
3   Plan making and development in terms of their procedures and outputs.
4   Outcomes and impacts of planning decisions in three key areas of planning – urban containment, settlement provision and landscape protection.
5   Hybrid evaluation by various methods and outlook for the future.

Or, put another way, as shown in more detail in Figure 2.11, Chapter 1 outlines the system as a prerequisite piece of description. Chapter 2 discusses how we can comprehend the system. Chapters 3 and 4 explain and interpret the system, and Chapter 5 evaluates and reviews the system.

The most innovative and pioneering aspect of the book is thus the attempt to provide in *one* book what has previously been attempted only in three types of book. First, books that merely outline the system; second, books that just describe planning processes and decisions; third, books that provide advanced theorisations of planning and its role in society but lack a practical appreciation of day-to-day planning reality. It is thus hoped that this book will be a landmark text in planning literature by providing in one book:

1   A thorough description of the system and its evolution.
2   A detailed account of its decision-making processes.
3   An analytical evaluation of its impact on land use and society.
4   An exhaustive evaluation of the system from a variety of theoretical standpoints.

1

How and Why the Planning System Evolved

The main purpose of the chapter is to provide a basic knowledge of the planning system, via its evolution, equivalent to a trainee doctor naming the parts of the body, as an essential prerequisite to the more rigorous discussions and evaluations in subsequent chapters.

The second purpose of this chapter is to provide an outline of why the planning system was developed and how it operates. The chapter takes an historical approach because legislation in Britain is strongly incremental, with one piece of legislation amending previous legislation. Accordingly, the reasons why one piece of planning legislation was amended provide a good evaluation of its strengths and weaknesses and also a guide to the goals of the revised legislation. Thus an historical approach provides the basis for an evaluation framework.

The chapter is divided into two main sections, which reflect the two main parts of the planning system, plan making and development control. Plan making sets out the goals of the planning system and crucially where the development of houses, shops and so on should be encouraged or discouraged. Development control is the process whereby applications to develop land are approved or refused in the context of the plans produced by plan making but also by taking into account other factors. Accordingly, the first part outlines the evolution of plan making and the second outlines development control procedures.

Learning objectives for Chapter 1

1. To learn what the planning system does as a key prerequisite to evaluation.
2. To use the evolution of the system as an early indication of where its strengths and weaknesses lie and its likely impacts.
3. To understand why it is divided into the two key functions of plan making and development control and the implications of the division.

HISTORICAL EVOLUTION OF PLAN MAKING

Historical evolution till 1947

Evolution from the 1850s to the 1930s

The first calls for planning began in the mid-nineteenth century when concerns over the disease-ridden squalor of Victorian industrial cities, for example cholera, led to the Public Health Act 1848. This Act merely sought to control standards in new development. It did little about the existing squalid towns nor did it concern itself with layout or environmental design. Nonetheless, it marked a recognition that it was in everybody's self-interest to intervene into the horrific conditions brought about by industrial towns and cities, because not only the workers were threatened by disease, but also the bosses.

The first of two reforming governments in the twentieth century, the Liberals in 1905–1910 (the other being Labour 1945–51) introduced the first major change with the Housing, Town Planning Act 1909. This Act enabled but did not require local authorities to make 'town planning schemes' for land that appeared likely to be needed for town planning purposes.

This voluntary principle highlighted the view of politicians at the time that the role of the state was to safeguard people from injustice, because progress came via private enterprise. Thus under the 1909 Act local authorities could intervene only if they felt that private builders would not provide acceptable housing or factories. If they did make a scheme the local authority could 'secure proper sanitary conditions, amenity and convenience in connection with the laying out and use of land and of any neighbouring land'.

Further progress was made in the First World War when the troops who suffered terribly in the trenches in France were kept going by the promise that they would return to a 'land fit for heroes'. The result was the Housing, Town Planning Act 1919. This made the drawing up of a town planning scheme compulsory for towns above 20,000 people. The main theme of the Act, however, was council housing, inspired by philanthropic enlightened company towns like Bournville in Birmingham (Cadbury's chocolate) and Port Sunlight in Liverpool (Lever Brothers' soap) in which it was argued that well housed workers were happier and more productive workers. Council housing was also inspired by the private New Towns movement created by the Town and Country Planning Association and its initial investment at Letchworth Garden City to the north of London.

In addition, the work of the founding fathers of planning, Ebenezer Howard, F.J. Osborne and Patrick Geddes, demonstrated how cities need not be dark satanic places but could be leafy green spaces if housing and jobs were separated by zoning. A crucial factor in this separation was rapid transport, first by suburban railways and then by bus and car. The 1920s thus saw the creation of a new urban form, the suburb. The work of Sir J. Tudor Walters was inspirational, as was the example of suburbs in the United States, as exemplified by Radburn, New Jersey, and the work of Frank Lloyd Wright, who created the antithesis of suburban greenery, Central Park in New York.

In particular, Tudor Walters in a key report argued:

> Do not let us set out to fill up all the little bits of spaces in the centres of our towns with badly planned small houses. Let us instead go right out into the suburbs of our towns and cities. Let us have belts of new housing schemes, planned and laid out on lines that are spacious and generous in their conception and execution.

In order to achieve Tudor Walters's vision the 1919 Act gave local authorities powers to

1 Limit building densities per acre.
2 Specify the percentage of a building plot to be a house or a garden.
3 Specify the character of buildings.
4 Determine the lines of new 'arterial' roads and buildings along these.
5 Specify the provision of open space.

The booming economy of the 1920s and the new mobility offered by buses and cars and expanding suburban railways created not only a new artefact, the suburbs, but also a new class, the middle class. This new class had new values, in particular a concern for the environment expressed ironically in a desire to create a microcosm of village England in their suburban plot with its Tudorbethan architecture and its carefully managed gardens. At the same time this new middle class used another new concept, leisure time and the mobility offered by the car, to seek recreation in the countryside. When they visited the countryside what they found there, however, was a loss of open countryside due to the 'zeitgeist' of Tudor Walters's vision plus the lax controls of the 1919 Act.

Particularly grubby developments were the 'plotlands' (Hardy and Ward, 1984) where individuals had hand built shack-style houses in the open countryside. The *cause célèbre* of the period was, however, a development called Peacehaven on the white cliffs of south-east England where the desecration of this symbolic landscape by a wasteful sprawl of houses created a national outcry against the lack of planning controls. Concerned by the way in which the countryside and the coast were being built over, the new middle classes were key players in the growth or formation of three of the core organisations involved in planning, as shown in Box 1.1. There is, however, a fundamental schism between two of these groups. The Council for the Preservation of Rural England has always campaigned for building to be restricted to towns where possible, while the Town and Country Planning Association, reflecting its roots in the New Towns movement, is broadly in favour of new settlements in the open countryside where appropriate.

Box 1.1   Key organisations involved in creating the planning system

The Town and Country Planning Association

The main pressure group. It grew from the New Towns movement in the 1920s and is still dominated by discussions about creating new settlements. It is also an advocate of more house building at lower densities.

The (Royal) Town Planning Institute

The professional organisation for planners, which sets standards. Membership is essential for virtually all planners. Most people now qualify as undergraduates or post graduates.

Campaign (Council) for the (Protection) (Preservation) of Rural England

A sectoral pressure group founded in the 1920s. It has changed its name several times to reflect changing attitudes to conservation. Though a small body it is very influential in promoting an anti-development ethos.

Together these organisations were instrumental in getting the Town and Country Planning Act 1932 on the statute book. This was the first Act to use the title 'Town and Country Planning' and to embrace all types of development land. The major difficulty with the Act was that it was based on 'zoning'. This meant that 'conforming uses' (i.e. land uses that conformed to the plan) were given permission, but refusals led to compensation. This was not only prohibitively expensive, but also induced speculative proposals to get compensation. In addition, too much land was zoned. The Act was thus too weak because landowning and building interests were able to water down the proposals made by the planning organisations. This was because the state was still seen as a body which prevented bad things, rather than promoting good things.

The catalyst of the Second World War

The economic depression of the 1930s began to change this view. In particular, the thinking of the famous Cambridge economist, John Maynard Keynes, and the example of the 'New Deal' in the United States, with massive new reservoirs and other public projects, being used to regenerate the economy introduced the concept of 'state intervention'. The really major change in thinking was, however, brought about by the Second World War, when the following factors were instrumental in a change of attitudes to politics:

1   The advent of the radio allowed the circulation of ideas to captive audiences of troops.
2   Mixing of classes in both factories and the armed forces.
3   Travel and vision of other forms of urban development, most notably in Europe.
4   Centralised planning for the war effort showed how socialism could work.
5   Earlier failure of capitalism in 1929, with the Wall Street Crash and subsequent economic depression of the 1930s followed by the perceived success of the socialist 'New Deal' in the United States, and communism in the Soviet Union.

In sum, this created a climate of opinion in favour of a more equal and also a more planned society. Thus the war spawned a series of reports, commissions and debates that produced the 'welfare state', the nationalised industries of coal, steel, transport and other public services, the setting up of the National Health Service, and the centralised planning of post-war Britain. The most important of these reports, named after their chairmen, were:

The Barlow report (1940) pointed out the strategic and geographic disadvantages of having too much growth in the south-east of England. In particular in terms of congestion and inflation in the south-east of England and the destruction of social capital elsewhere, notably in the areas of former heavy industry. The solution was seen to be regional policy, overseen by a central planning authority and backed up by dispersal to garden cities, satellite towns and trading estates. The Barlow report was immensely influential and its twin policies of restricting development in the South East and encouraging it elsewhere have been the two core policies of post-war regional policy.

The Scott report (1942) on 'Land Use in Rural Areas' considered the run-down nature of Britain's farmland that had suffered from nearly 100 years of uninterrupted imports from North America and Australasia. In addition agriculture had not been supported by subsidies and so it was unprofitable in the 1930s. As a result land prices were low and this exacerbated the rapid suburbanisation of the 1930s as land for housing could be bought very cheaply and easily. But the war cut off food imports that had accounted for about 70% of all food consumed in the 1930s. Thus Britain suffered severe food rationing during the war as a result of the German submarine blockade. The need to retain agricultural land after the war was therefore deemed to be paramount by the Scott report. It therefore recommended that: (1) farmland would need to be protected from urban growth far more vigorously and (2) agriculture and forestry should not be fettered by planning controls. In spite of a minority report by Dennison this led to the fatal divorce of most countryside issues in the inappropriately named Town and Country Planning Act in 1947. The main purpose of the Act was indeed to contain urban growth and protect the countryside from new building.

The Uthwatt report (1942) on 'Betterment and Compensation' made the very important point that planning can work only if a refusal of planning permission does not lead to the payment of compensation. The corollary of how to deal with 'betterment', the huge rise in land value given by the grant of planning permission, is not so easily dealt with and 'betterment' remains a problematic area for planning. In contrast compensation on refusal was limited to pre-existing plans by private developers (i.e. before these measures were proposed) and were limited to 1939 values. In the early 1950s a once-and-for-all payment of around £380 million was made – probably around £10 billion in late 1990s money or around 3% of annual government spending, a huge amount but a necessary prerequisite for effective planning.

These three reports were matched by other famous reports, notably the Beveridge report on social services, pensions and the NHS. They all proposed a massive change from a private enterprise system to a socialist system which is usually known as the welfare state, or, as Margaret Thatcher later derided it, the 'nanny state' in the 1980s. It is hard to comprehend now how socialist a country Britain was in the late 1940s but it is fundamentally important to understand that planning was founded as a socialist enterprise. Nonetheless, the vested interests of private enterprise fought tooth and nail against the imposition of socialism.

In spite of their opposition socialist legislation began to unfurl from the three reports. First, the Town and Country Planning Act 1944 allowed wholesale purchase of all land needing comprehensive development as a result of war damage at 1939 values. The faces of many city centres have been massively influenced by this Act, notably Coventry and Plymouth. The Act also introduced compulsory purchase of land where necessary to secure properly planned development. This power has been extensively used, notably to acquire land for transport facilities.

The stage was now set for the introduction of proper planning, and the entry of a Labour government in 1945 with a majority of 100 seats provided the actors. However, the immense task of developing the 'welfare state' forced the Labour Party to dissipate its energy into *four* environmental Acts, with devolved/delegated powers rather than centralised ones. The passage of four Acts, however, meant that the sum of the parts was less than the whole would have been. The four main Acts were:

1   *The New Towns Act 1946*, which allowed the construction of new towns.
2   *The Agriculture Act 1947*, which supported agriculture in its post-war expansion and thus made farmland desirable for farming again, rather than for development as in the 1930s.
3   *The National Parks and Access to the Countryside Act 1949*, which allowed the creation of National Parks and Areas of Outstanding Natural Beauty and encouraged recreational access to these areas.
4   *The Town and Country Planning Act 1947*, which with modifications contains the essentials of the system today.

Plan-making procedures from 1947 to 2004

There have been two major periods of plan making:

1   *1947–68, with a long-term hangover till the mid-1980s.* Development plans based on the 1947 Act, with emphasis on the local authority as the developer. Plans were not based in the reality of a rapidly changing Britain and had to be replaced.
2)   *1968 (1971 Act) till 2003, but with revisions in the 1990s.* Plans changed to become guides

for development control based on a strategic 'Structure' Plan and a detailed 'Local' Plan. An overarching regional tier was removed in the 1980s but returned in the 1990s and looks set to become dominant in the twenty-first century.

Thus this section is divided into three: (1) the plan-making process in the 1950s and 1960s; (2) the plan-making process in the 1970s and 1980s; (3) modifications of the process in the 1990s. Changes to the system in 2004 are outlined at the end of Chapter 5.

The plan-making process in the 1950s and 1960s

Section 5 of the Town and Country Planning Act 1947 laid down that:

> Every local planning authority shall carry out a survey of their area . . . and shall submit to the Minister a report of the survey . . . together with a plan (hereinafter called a 'development plan') indicating the manner in which they propose that land in their area should be used [over 20 years] . . . and the stages by which any such development should be carried out.

We can thus see here the classical cyclical approach to planning first advocated by the Edinburgh polymath Patrick Geddes at the beginning of the twentieth century, namely: (1) survey, (2) plan preparation, (3) approval, (4) implementation, (5) monitoring, (6) review.

A key feature of the 1947 system was that local authorities were expected to be the developers, building most if not all the houses, schools, factories and shops. The development plan thus split land into four types of area:

1   White land (farmland where following the Scott report development would be resisted).
2   Existing use areas (where only minor change was expected).
3   Development and redevelopment areas (areas where major changes were expected).
4   Redevelopment of war damage.

Public participation was in the form of a 'public inquiry' and approval by the Minister. But this was very much a rubber-stamping exercise. The general public were told what was good for them and were not asked their views till late in the day. The next stage – implementation – was done in two ways.

1   *Positive* planning by purchase and development of land by the local authority using 'Designated Land' (land to be compulsorily purchased) and 'Programme Maps' (stages by which land would be developed). It was expected that the local state would build the vast majority of houses, shops, work units and public buildings like hospitals. The apogee of the system was the master plans for the new towns which were built between 1946 and 1980 by the central state. At the local level local authorities constructed huge council housing estates under these provisions in the 1950s.
2   *Negative* planning was to be by development control of applications for the private development of land. But this was meant to be for a small residual amount of development outside the public sector.

The final stage, the review of how the plan was getting on was supposed to take place every five years. In essence the 1947 system contained three key features:

1   It was based on the local authority as developer.
2   It was predicated on large-scale developments like housing and employment estates.
3   It used the map as the main form of analysis/presentation.

The 1947 system was replaced gradually from 1968 onwards, for six basic reasons:

1   It overemphasised land use at the expense of other issues, notably transport.
2   Society changed its views. Socialism lost its appeal, and individual freedom became the 'zeitgeist' of the 1960s.
3   Economic and population growth was faster than expected and so the slowly produced plans became chronically out of date and out of touch. Most areas did not have a plan till well into the 1950s and some not till the 1960s.
4   The plans were too rigid, too simple and too long-term and based too much on a map.
5   Little if any social/economic context; too much physical design, not enough on how people would live.
6   Change of government in 1951 from socialist Labour to old-style conservatism based on patrician values but with several memorable election slogans: 'Set the people free' (1951) 'You've never had it so good' (1959).

In particular, new population forecasts in 1965 predicted a dramatic rise in population compared to previous forecasts as the birth rate accelerated and thus the need for a more dynamic style of planning. In addition, the changing nature of the housing market was fundamental to change, as people began to see housing as an investment not merely as a place to live. For example, in the 1950s the Conservatives under the Housing Minister, Harold Macmillan, set a target of 400,000 new homes a year, a figure only once ever achieved in the early 1960s, but ironically achieved in the 1950s by now derided council house building. Local authority (council) houses dominated construction in the early 1950s, with nearly 200,000 such houses being built in 1955, but by the late 1950s private houses took over and from 1963 private houses accounted for about 60% of all new housing. This change meant that plans became guides for development control rather than as intended development plans for local authorities. Thus the system was doomed, since its *raison d'être* had been destroyed.

Accordingly, the system was reviewed in the 1960s by three reports, as shown in Box 1.2. The PAG report (Housing and Local Government, 1965) emphasised the need for a different type of plan that would be more strategic, more related to motor-car use, given that car numbers were rapidly growing, and more useful for the control of development rather than its instigation. The Skeffington report (Housing and Local Government, 1969) reflected the idealism of the 1960s and the move away from a patrician Tory government, towards greater social inclusion. The Redcliffe Maud report (Cmnd 4039, 1969) reflected both reports with an emphasis on a modern technocratic Britain based on the new concept of city regions dominated by motorised travel and the rejection of a Tory past based on shire counties.

However, in 1970 the incoming Conservative government for ideological and sentimental reasons legislated in 1972 and 1974 for a two-tier system based on counties and districts. Appropriately, the new system shown in Figure 1.1 came into force on 1 April 1974. The main problems were as follows. First, the metropolitan areas like the West Midlands and South Yorkshire were too small to be able to plan for any expansion. Second, counties were out of date and did not fit the new patterns of life based on city regions. Third, the district boundaries were absurd, for example around Exeter in Devon the boundaries of several local authorities overlap. Fourth, too many large towns and even cities were downgraded in status, for example Bristol lost its century-old independence and became part of the artificial new county of Avon.

Box 1.2 Summary of three key reports in the 1960s

Planning Advisory Group (PAG), *The future of development plans* **(1965)**

This report highlighted the six problems already discussed and concluded that there would have to be a *radical change in the form and content of development plans.* The PAG thus recommended a different type of plan:

1 *Statements* of policy and broad principles only illustrated by diagrams and by a *structure* map for sub-regional areas.
2 *Local Plans* for smaller areas in order to put in detail.

Two types of plan were proposed:

1 *Urban areas* above 50,000 people. Broad zones and transport.
2 *Rural areas.* Broad pattern of people, employment, major communications, main policies for conservation, recreation.

Skeffington report on public participation (1969)

The 1960s were a period of student protest, notably over the Vietnam War, and so *participation* was a buzz word. Accordingly, Skeffington, a Labour MP was asked to show how the public could participate in planning not just when the plan was published but at every stage in a vibrant exchange of ideas about the future of a community.

Redcliffe Maud report on local government (1969)

Redcliff Maud an Oxford don led a team of experts who produced a massive, well researched and logical report on modernising local government, as had been imposed on post-war Germany. This would have been the first major reform since Victorian times.

Unitary authorities based on the concept of the *City Region* in order to provide the ideal basis for the plan-making system centred on *Structure Plans* proposed by the PAG were thus proposed by Redcliffe Maud.

Box 1.3 Split between planning functions from 1974 onwards

Function	County	District
Structure Plan	•	○
Local Plan	○	•
Development control	Minimal	•
Transport	•	Minimal

• marks where the function is carried out. *Minimal* marks where the function is minimally carried out. ○ marks where there is no formal function though there may be informal links.

Figure 1.1 New administrative areas as at 1 April 1974. *Source*: Gilg (1978)

The worst problem was that the new planning system based on the PAG report and legislated for in the Town and Country Planning Act 1968 (later consolidated into the 1971 Town and Country Act) had assumed that the plans would be prepared by city region authorities as proposed by the Redcliffe Maud report. This produced a split of functions between county and districts as shown in Box 1.3.

This split caused three main problems. First, Structure Plans were supposed to provide the framework for Local Plans but were prepared by two different authorities, counties and districts, thus opening up the possibility of conflict, especially if two different political parties were involved. Second, plans were to be implemented by development control but counties were given minimal development control powers. Third, transport had become a major feature of planning by the 1970s but counties were given most of the powers over transport whereas most day-to-day planning was done by the districts.

The plan-making process in the 1970s and 1980s

The 1970s thus witnessed the creation of a radically new system of planning based on two tiers of administration and two tiers of plan and lip service to public participation, which carried on through the 1980s. The two key elements of the system, Structure and Local Plans, are now considered in turn.

Structure Plans

Advice on how to prepare Structure Plans was contained in two main documents in the 1970s and 1980s: a manual on Development Plans and a DOE circular. The manual produced by the Ministry of Housing and Local Government (1970) set out detailed advice in a thick A4 style format, but also in brief as shown in Box 1.4. The 1970 manual influenced the first generation of Structure Plans in the 1970s, but was then replaced in 1984 by a much simpler and shorter piece of advice, which built on the experience of the 1970s but also reflected the minimalist approach to planning of the Tories in the 1980s.

The new advice contained in DOE Circular 22/84 (DOE, 1984) set out a sequence of steps for producing structure plans, as shown in a simplified form in Figure 1.2. The process started with a survey of those matters expected to affect the development of the area, notably: land use, population and communications. The survey, as well as indicating existing problems, was to be used to make projections from which a 'Report of Survey' should outline a number of policy options backed up by an 'Explanatory Memorandum'.

These options were expected to take into account: current policies for the region as a whole, the resources likely to be available and the opinions of the public. With regard to public participation, local planning authorities were told to undertake three tasks: provide adequate publicity about the plan; make people aware that they could participate and how, and provide copies of the plan.

After this process of policy formulation, policy choice and consultation, which might take several years, the plan had to be submitted to an 'Examination in Public' based on a document called the 'Written Statement' which set out the chosen policies. At the heart of the 'Written Statement' there was a 'Key Diagram' which set out the main policies for development in diagrammatic form. Individual chapters dealt with subject issues like housing and contained policies, which were numbered and lettered for reference purposes, for example Housing Policy 4 (HP4), 'Normally new housing will not be permitted in the open countryside unless there is a proven need e.g. to house an agricultural worker'.

The 'Examination in Public' (EIP) was a series of cross-examinations conducted between legal professionals and expert witnesses in front of a three-person panel as shown in Figure 1.3. From Figure 1.3 it is clear that the key players closest to the panel are the House Builders' Federation and the county council. In contrast the CPRE, residents' groups and mere observers are kept at arm's length. The EIP is normally structured by topics. A typical topic could be housing issues broken down into sub-topics, for example:

forecasts of population change/housing need, and location of new housing. At the debate different groups took different positions, for example residents' groups usually attacked the forecasts as too high while housebuilders' groups attacked the forecasts as too low or in the wrong place.

Box 1.4 Functions of Structure and Local Plans as set out in *Manual on Development Plans: Form and Content* (1970)

Structure Plans

1. Interpreting national and regional policies.
2. Establishing aims, policies and general proposals.
3. Providing framework for Local Plans.
4. Indicating Action Areas.
5. Providing guidance for development control.
6. Bringing before the Minister and the public the main planning issues and decisions.

Typical subjects: Population, Employment, Resources, Housing, Industry and commerce, Transport, Shopping, Education, Social and community services, Recreation and leisure, Conservation, Townscape and landscape, Utility services, other subjects (e.g. minerals).

Local Plans

1. Applying the strategy of the Structure Plan in detail.
2. Providing a detailed basis for development control.
3. Providing a basis for co-ordinating decisions.
4. Bringing detailed planning issues and decisions before the public.

District Plans

For comprehensive planning of relatively extensive areas, e.g. central area of a large town.

Action Area Plans

For the comprehensive planning of areas selected for intensive change by improvement, redevelopment or new development, e.g. the improvement of a run-down residential area.

Subject Plans

For the detailed treatment of a particular planning aspect, e.g. minerals, recreation.

In practice the EIP was centred on a few topics and dominated by established interest groups, notably those with a financial interest. It took place over several weeks or even months, and attracted little media interest. It did not follow the ideal model proposed by Skeffington in the 1960s except in terms of lip service. Thus it tended to be a process carried out by consenting adults in private.

At the end of the EIP a report was written which went to the Secretary of State for the Environment. The Secretary of State normally did one or both of two things. First, revise forecasts for say housing, shopping and population. Second, delete policies said to be *ultra vires* (i.e. outside the law). For example, in the 1980s the Lake District planning authority wished to include a policy to allow new house building only for local people, but this was deleted

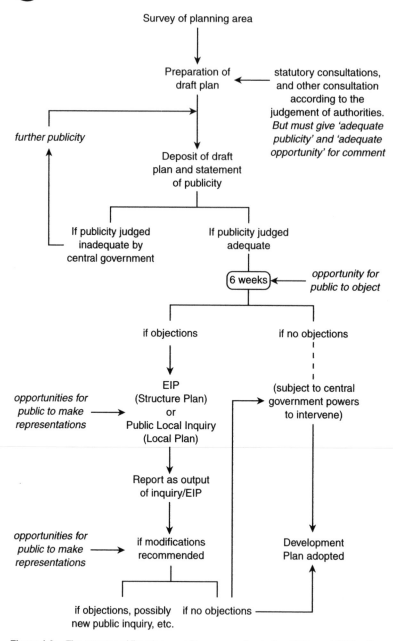

Figure 1.2   The process of Development Plan preparation in the 1980s and 1990s. *Source*: Thomas (1996)

because it was outside planning law. This is because planning is restricted to land use planning and cannot be used for social issues.

Table 1.1 shows that the scale of changes made in terms of housing provision was between 5% and 10% in most cases. There are, however, strong regional variations, with most plans in the South and the Midlands being revised upwards, but provision in the North being either

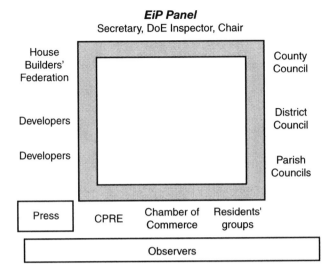

Figure 1.3 Typical arrangements at an Examination in Public. *Source*: adapted from Murdoch and Abram (2002)

kept the same or actually reduced, as in North Yorkshire. The table also shows that even in areas of restraint like Cumbria, which then included parts of the Lake District National Park, substantial numbers of new houses are provided for by the planning process.

However, as Figure 1.4 shows, Structure Plan authorities have also made substantial changes to their housing provision between Structure Plans as the review process got going in the early 1990s. For example, some counties, for example Warwickshire and Northamptonshire had increased their provision by around 20%, while others like Surrey and Durham had reduced their provision by as much as 25%. No clear spatial pattern emerges from the figure but there does appear to be a tendency for areas around London and in the North to have reduced their provision and areas in the Midlands and the South West to have increased their provision.

Table 1.1 Examples of changes made by the Secretary of State of the Environment to Structure Plans after the EIP

Structure Plan	Original provision	Revised provision
Somerset	34,750	36,750
Bedfordshire	32,800	38,750
Suffolk	35,500	42,000
Warwickshire	25,300	28,950
Leicestershire	53,100	55,000
North Yorkshire	45,800	42,000
Humberside	45,000	45,000
Cumbria	23,600	23,600

Source: Journal of Planning and Environment Law, October 1988, p. 682

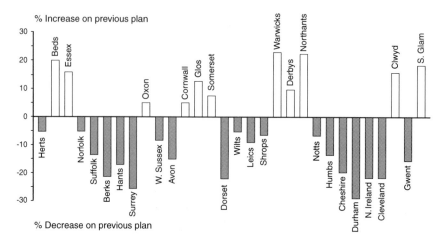

Figure 1.4 Changes in housing provision between Structure Plans. *Source:* adapted from Coates (1992)

Implementation of the plan was via development control. But there is a major problem here. The Structure Plans were prepared by the counties but the districts were primarily to carry out development control. Why should the districts take any notice of the Structure Plan, especially if they are of a different political persuasion? There have been and are four safeguards against this. First, between 1974 and 1980 the counties could 'call in' applications to make the decision. In 1980 the Local Government Planning and Land Act 1980 replaced 'call in' with a clause asking district councils to 'seek the general objectives of the Structure Plan'. Between 1974 and the late 1990s a 'Certificate of Conformity' with the Structure Plan had to be given by county councils when district councils produced their Local Plans. Finally from the mid-1990s many local authorities became unitary authorities, thus eliminating the difficulty in some areas. Most of these safeguards are weak and flawed and the net result is that the implementation of Structure Plans has been out of the hands of those who produce them.

The final part of the process is 'monitoring' and 'review'. This emphasises the cyclical and incremental nature of the process, with one Structure Plan being followed by a modified version on a rolling five- to ten-year programme. The intention is to produce a framework for planning that is firm but also flexible enough to accommodate change. Figure 1.5 emphasises the negotiative aspect of the process as well as providing a useful summary.

Local Plans

Local Plans were meant to be more like the 1947-style Development Plan, with a map being the centrepiece rather than the diagram at the heart of the Structure Plan. The three key features of a Local Plan were first to put detail on to Structure Plans at the local level, second to provide a shorter time horizon, and third to provide a guide for development control. As with Structure Plans advice on their preparation was provided by the 1970 manual and the 1984 circular.

Local Plans could take three forms: District Plans, Action Area Plans and Subject Plans. District Plans were the norm and were concerned with slow, piecemeal change over a wide (normally a district council or town area). Action Area Plans were for more rapid

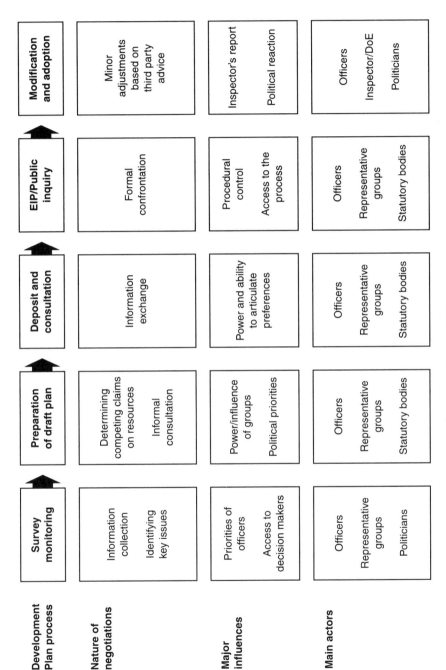

Development Plan process	Survey monitoring	Preparation of draft plan	Deposit and consultation	EIP/Public inquiry	Modification and adoption
Nature of negotiations	Information collection Identifying key issues	Determining competing claims on resources Informal consultation	Information exchange	Formal confrontation	Minor adjustments based on third party advice
Major influences	Priorities of officers Access to decision makers	Power/influence of groups Political priorities	Power and ability to articulate preferences	Procedural control Access to the process	Inspector's report Political reaction
Main actors	Officers Representative groups Politicians	Officers Representative groups Statutory bodies	Officers Representative groups Statutory bodies	Officers Representative groups Statutory bodies	Officers Inspector/DoE Politicians

Figure 1.5 Negotiations in the Development Plan process. *Source:* Claydon (1996)

(re)development, rehabilitation, improvement and were often used for city centre plans. They were often also the most successful, since they had clear aims, short time horizons (five years typically) and greater control, e.g. land assembly. Subject Plans were the third type of Local Plan and were for individual topics, for example Green Belts, Minerals and Landscape.

Key policies in the Local Plan identified enough land for 10 years' housing development and enough land for minerals for 15 years. Policies also ensured that no more than the essential minimum of farmland was to be taken and that land of a higher quality was not to be taken when less good land was available.

Local Plans were prepared in a similar way to Structure Plans but within the framework of the Structure Plan. The big problem with Local Plans was their uneasy relationship with Structure Plans. In more detail people did not foresee/were unaware/failed to win their case at the EIP. The Structure Plan predated and to some extent settled the framework for the Local Plan, and the Structure Plan once approved was a legal document. Once the local implications of the Structure Plan become apparent at the Local Plan stage, protesters realised too late that they could not protest about the overall thrust of the plan, but only its detail. Thus they became upset, and felt they had been cheated. In practice, objections were allowed, and indeed Local Plans did contradict Structure Plans, but this discredited the basic logic of a two-tier system.

As with Structure Plans the key public debate was at the 'Public Local Inquiry' but this was less adversarial and less formal than the EIP and conducted by a Planning Inspector rather than by a panel (see section in Chapter 3 for more detail on Planning Inspectors). However, cynical commentators say that this was a cosmetic public relations exercise to legitimise conspiratorial decisions made behind closed doors by a few influential people. A key difference, however, lay with the resulting report by the inspector. This did not go to the Secretary of State for a final verdict but instead to the local planning authority, who could act as they saw fit on the report's recommendations.

As Table 1.2 shows, the initial local plans were produced very slowly, notably at the preparation stage, with plans typically taking three years to be produced and then two further years to come into operation. Table 1.3 shows that the process of Local Plan approval continued to be a slow process in the 1990s. Even the two plans in their review stages when it could be presumed that most of the major issues had been agreed upon took four and six years. In the worst cases some plans had not been adopted six, eight or even twelve years after the draft had been published. The main fault seems to lie within the local authorities, for the inspectors' reports were on average issued only a few months after the completion of the inquiries.

Implementation took place by a combination of activities. First, private enterprise was 'encouraged' by the plan to develop in the locations favoured by the plan and 'discouraged' (prevented) from developing elsewhere by development control. Second, compulsory purchase of land was used where necessary to secure land for development. Third, public works, for example new roads, may be built as a catalyst for development. Fourth, partnership schemes between public/private enterprise: in recent years planners have been encouraged to see their role as partner building by involving all stakeholders in a form of communitarian activity.

The process as before was completed by monitoring and review and as before this was in the form of a rolling programme with incremental policy changes. The overall system as it operated in 1980 and then till the early 1990s is shown in Figure 1.6 with the exception of modifications to planning in the big cities which were enacted in the mid-1980s.

Table 1.2 Stages and timing in preparation of the first generation of Local Plans

Stage	Timing (months)	Running total
Preparation and publicity for local plan brief	1.5	1.5
Preparation of Report of Survey and public participation	18.0	19.5
Preparation of draft local plan and public participation	12.0	31.5
Production of 'final' local plan	2.0	33.5
Certification of conformity with Structure Plan	1.0	34.5
Deposit	1.5	36.0
Public Local Inquiry	7.0	43.0
Inspector's report received	2.5	45.5
Modification on Deposit	4.0	49.5
Adoption of local plan	2.5	52.0

Source: Healey (1983)

Table 1.3 Examples of Local Plan preparation and modifications

Stage	Ashford Local Plan second Review	Maldon Local Plan Review	Southern Derby Local Plan	Gloucester Local Plan	Combe Martin (Devon) Local Plan	Carlisle Local Plan
Draft	October 1990	April 1990	May 1990	September 1988	May 1990	November 1984
Deposit	June 1991	November 1990	July 1991	September 1991	January 1991	September 1985
Completion of inquiry	February 1992	November 1991	July 1992	September 1993	October 1991	January 1990
Inspector's report	October 1992	April 1992	December 1992	June 1994	March 1992	March 1990
Modifications published	April 1993	September 1993	May 1993	February 1995	April 1992	July 1990
Modifications inquiry	December 1993	June 1994	November 1993	March 1996	May 1992	May 1991
Inspector's report	January 1994	September 1994	December 1993	?	?	July 1991
Adoption	July 1994	August 1996	?	?	January 1994	?

? Denotes where the authors were not sure about the status of the plan at this stage.
Source: Crow *et al.* (1997)

In the big cities (the metropolitan authorities of Greater London, West Midlands, Greater Manchester, Merseyside, South Yorkshire and West Yorkshire) major changes were made in 1985 to the process of plan making as a consequence of the Local Government Act 1985. The changes arose from an ideological divide between Labour authorities at the local level and the Tory Party at the national level, at a time when the Labour Party was far more left-wing than it became under Tony Blair. In particular, Ken Livingstone in London had subsidised London Transport fares, at the expense of suburban commuters. In order to get rid of this ideological divide the Tory government scrapped the big metropolitan authorities set up in 1974.

They were replaced by a collection of District Councils with no superior or co-ordinating authority. This necessitated a new type of plan, 'Unitary Development Plans', in the former metropolitan areas. These consisted basically of a two-tier plan: Part I, overall policies (comparable to a Structure Plan), and Part II, detailed and site-specific policies (comparable to a Local Plan). The evolution of strategic planning guidance for both counties and metropolitan areas between 1980 and 1995 is shown in Figure 1.6, which also shows the situation in London. London was ironically without a strategic plan for many years until a draft plan for London was published in 2002.

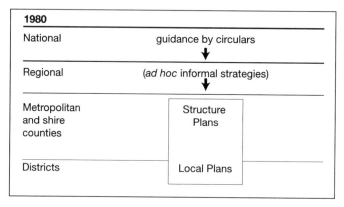

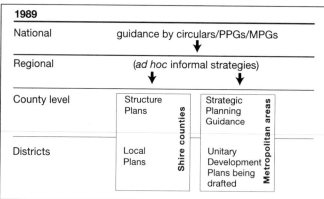

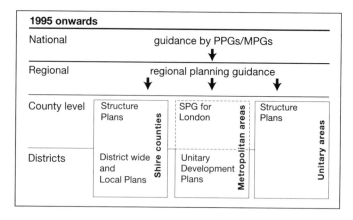

Figure 1.6 Strategic planning frameworks in England, 1980–95. *Source:* adapted from Thomas and Roberts (2000)

Modifications of the plan-making process in the 1990s

The advice in DOE Circular 22/84 was modified and eventually replaced by a series of new policy advice documents, Planning Policy Guidance Notes (PPGs) as shown in Box 1.5. In terms of plan making the most important were PPG1 (General Policy and Principles), published first in 1988 and revised in 1992 and 1997, and PPG12 (Development Plans and Regional Planning Guidance), published first in 1988 and revised in 1992 and 1999. The first PPG12 (Department of the Environment, 1988) then entitled *Structure and Local Plans*, set out a list of eight key structure plan topics:

Box 1.5   Policy guidance to local planning authorities

Planning policy guidance in England		National Planning Policy Guidance Scotland	
PPG1	General Policy and Principles	NPPG1	The Planning System
PPG2	Green Belts		
PPG3	Housing	NPPG3	Land for Housing
PPG4	Industrial Development	NPPG2	Business and Industry
PPG5	Simplified Planning Zones		
PPG6	Town Centres and Retailing	NPPG8	Town Centres and Retailing
PPG7	The Countryside	NPPG15	Rural Development
PPG8	Telecommunications	NPPG19	Radio telecommunications
PPG9	Nature Conservation	NPPG14	Natural Heritage
PPG10	Waste Management	NPPG10	Waste Management
PPG11	Regional Planning		
PPG12	Development Plans		
PPG13	Transport	NPPG17	Transport
PPG14	Unstable Land		
PPG15	Historic Environments	NPPG18	Historic Environments
PPG16	Archaeology	NPPG5	Archaeology and Planning
PPG17	Sport and Recreation	NPPG11	Sport, Physical Recreation and Open Space
PPG18	Enforcement		
PPG19	Outdoor Advertisements		
PPG20	Coastal Planning	NPPG13	Coastal Planning
PPG21	Tourism		
PPG22	Renewable Energy	NPPG6	Renewable energy developments
PPG23	Pollution Control		
PPG24	Noise		
PPG25	Flood Risks	NPPG7	Flooding

A series of Minerals Planning Guidance Notes is issued for England.
Policy Guidance for Wales has been issued as a single document, supplemented by Technical Advice Notes and Minerals Planning Policy Wales.
In Scotland, National Planning Policy Guidance is supplemented by more detailed Planning Advice Notes.

Other subjects covered in Scottish Guidance

NPPG4 Land for mineral working
NPPG9 Roadside facilities
NPPG12 Skiing developments
NPPG16 Opencast coal

Source: Royal Commission (2002)

1 New housing, including figures for housing provision in each district and where appropriate the role envisaged for new settlements.
2 Green Belts and conservation in town and country.
3 The rural economy.
4 Major industrial, business, retail and other employment-generating development.
5 Strategic transport and highway facilities.
6 Mineral working (including the disposal of mineral waste) and protection of mineral resources.
7 Waste disposal, land reclamation and reuse.
8 Tourism, leisure and recreation.

The advice in the 1992 revision of PPG12 followed on from revised legislation in the Planning and Compensation Act 1991, which gave much greater importance to development plan policies in deciding planning applications, as explained more fully on page 36.

Other major changes emanating from the Planning and Compensation Act 1991 were the introduction of authority-wide Local Plans, and the introduction of Minerals and Waste Local Plans and National Park Plans for both Land Use and Minerals. The main procedures remained the same but the government became less involved in terms of plan approval; in particular, Structure Plans were no longer normally subject to central government approval. Finally, in a reflection of greater commitment to sustainability, plans had to be subjected to an Environmental Appraisal. According to Tewdwr-Jones (1999) the 1991 Act had three main effects for plan making. First, it removed the discretion to prepare a plan or not with a duty to prepare a district-wide plan. Second, it continued professional discretion in terms of what issues to include in a plan and how to present them, and third, discretion remained with regard to interpreting legislation but policy discretion was reduced by the diktats of RPG spatial and housing targets. This was borne out by Baker (1999) who in a survey of 263 local planning authorities found that central government had been keen to intervene to ensure that PPGs were followed and that in particular restrictive housing policies had been watered down. The main changes in the early 1990s were:

1 Structure Plans retained but no longer subject to central government approval in common with Local Plans. But reserve powers for central government call-in.
2 District-wide Local Plans outside the former metropolitan areas.
3 Minerals Local Plans to be prepared by districts.
4 National Park Plans for Land Use and Minerals to be prepared by new National Park authorities.
5 Introduction of a so-called 'plan-led' system, with more power for plans in development control decision making (explained more fully in Chapter 3).
6 Plans to be subject to an Environmental Appraisal.

Plan making was further modified in the mid-1990s by a chaotic reform of local government in England into a mix of unitary/county–district authorities, as shown in Figure 1.7. In the two-tier areas the system remained much the same with County Structure Plans, but with District-wide Local Plans replacing Local Plans, which had till then tended to focus on specific towns. In the new unitary areas Authority-wide Structure Plans were

Figure 1.7 Planning authorities in the United Kingdom. *Source*: adapted from Cullingworth and Nadin (2002)

to be prepared. In the former metropolitan unitary areas the Unitary Development Plans of the 1980s continued. The overall effect of the changes by 2002 is shown in Box 1.6 and Table 1.4.

Box 1.6 The scope of Development Plans in England

The Town and Country Planning Act 1990 requires local planning authorities to consider:

1 The conservation of the natural beauty and amenity of the land.
2 The improvement of the physical environment.
3 The management of traffic.

In addition subjects that may be considered in Structure Plans are:

4 Housing, including figures for additional housing requirements in each district, and targets for development on greenfield and brownfield sites.
5 Green belts.
6 The conservation and improvement of the natural and the built environment.
7 The economy of the area, including major industrial, business, retail and other employment-generating and wealth-creating development; a transport and land use strategy and the provision of strategic transport facilities, including highways, and other transport requirements.
8 Mineral working (including disposal of mineral waste) and protection of mineral resources; waste treatment and disposal, land reclamation and reuse.
9 Tourism, leisure and recreation.
10 Energy generation, including renewable energy.

In relation to sustainable development PPG12 suggests 'other issues that may be addressed in plans, either as land use policies or as considerations which influence policies in the plan'. These include:

1 *Environmental considerations*: energy, air quality, water quality, noise and light pollution, biodiversity, habitats, landscape quality, the character and vitality of town centres, tree and hedgerow protection and planting, revitalisation of urban areas, conservation of the built and archaeological heritage, coastal protection, flood prevention, land drainage, groundwater resources, environmental impacts of waste and minerals operations, unstable land.
2 *Economic growth and employment*: revitalisation and broadening of the local economy and employment opportunities, encouraging industrial and commercial development, types of economic development; generally to take account of the needs of business while ensuring that proposals are realistic.
3 *Social progress*: impact of planning policies on different groups, social exclusion, affordable housing, crime prevention, sport, leisure and informal recreation, provision for schools and higher education, places of worship, prisons and other community facilities, accommodation for gypsies; but to 'limit the plan content to social considerations that are relevant to land use policies'.

Source: Cullingworth and Nadin (2002)

In the late 1990s local authorities were also given a duty to produce 'Community Strategies' in order to promote or improve the economic, social and environmental well-being of their areas. The intention is that the strategies should define and clarify overall priorities by establishing a small number of key policy objectives for the area, which would then be reflected in Land Use Plans.

Changes in the mid-1990s in Scotland and Wales were more rationally consistent and were based on the creation of unitary authorities. In Scotland, unitary authorities were to produce a Structure Plan and a Local Plan. In Wales authorities were to prepare a Unitary Development Plan, in which Part 1 is equivalent to a Structure Plan, and Part 2 to a Local

Table 1.4 Structure of local government in 2002

England outside London
In the six metropolitan areas (Tyne and Wear, West Midlands, Merseyside, Greater Manchester, West Yorkshire and South Yorkshire) thirty-six metropolitan district councils were formed in 1974 and 1985, most of them with a population of over 200,000. There are forty-five unitary authorities established between 1995 and 1998 most of which have a population of between 100,000 and 300,000. Elsewhere the 1974 structure remains, with thirty-four county councils and 238 district councils with populations in the range between 60,000 and 100,000. There were no plans in 2002 for further reforms. There are also a range of parish councils with limited planning functions.

London
The Greater London Authority was formed in 2000. It has a directly elected Mayor and Assembly. Thirty-two London boroughs and the City of London carry out local planning functions.

Scotland
A single tier of government was set up in 1996 with twenty-nine unitary authorities. These replaced the nine regional councils and fifty-three district councils set up in 1975. The three Islands councils remain. Over 1,000 community councils express the views of communities.

Wales
In 1996 twenty-two unitary authorities replaced the eight county councils and thirty-seven district councils set up in 1974. Seven hundred community councils act like English parish councils.

Source: Royal Commission (2002)

Plan. However, in all three countries not many of the authorities really matched the pattern of life in their areas and until planning authorities reflect the daily dynamics of city regions planning will always be trying to fit a square pin into a round hole.

The renaissance of regional planning

A more positive development in the 1990s was the comeback made by regional planning after a brief spell in the 1970s when Regional Economic Planning Councils had laid out broad frameworks for development. These 1970s plans were bids for government investment, e.g. the South West Economic Planning Council successfully got the M5 extended 40 miles further into the South West from Bridgwater to Exeter. They also set the regional context for regional housing allocations for Structure Plans, but Mrs Thatcher scrapped them soon after she came to power in 1979.

However, regional economic policy remained, but with funding increasingly from the European Union. The areas qualifying for assistance have remained depressingly similar and reflect the failure of the long-standing planning policy of resisting development in the South East in an attempt to deflect it to the North of England, Wales and Scotland. The European Union also has a non-statutory European Spatial Development Perspective and Euro-regions are becoming increasingly important. In the late 1990s Regional Development Agencies were set up with the task of preparing Regional Economic Strategies. In an empirical analysis of the process Allmendinger and Tewdwr-Jones (2000a) note that most commentaries have concentrated on the apparent 'hollowing out' of the state as devolution and regionalisation took root in the 1990s. Their survey of stakeholders, however, found concern over whether the new forums would be able to develop the capacity to harness both economic development and planning policy processes together.

Regional Land Use Plans were reintroduced early in the 1990s when the Tories realised that planning was not quite the socialist devil they had thought it to be in the 1980s. At first the plans were rather *ad hoc* and were prepared by groups of authorities. By the mid-1990s the process had become more formal and the plans became documents approved by the government as Regional Planning Guidance (RPG). All the local authorities in each region set up a regional planning body and they prepare a draft strategy in consultation with the government office for the region. After public consultation a revised plan is published and this is examined at a public inquiry (Gilg, 2000) which usually takes about four weeks. The plan is then approved with modifications by the Secretary of State. Key elements of the system are outlined below:

1 The draft RPG is produced by the regional planning body, consisting largely of the local authorities acting collectively.
2 The draft RPG undergoes public examination conducted by a panel of experts appointed by the Secretary of State.
3 The public examination panel presents a panel report to the Secretary of State with recommended changes to the draft RPG.
4 The Secretary of State publishes proposed changes to the draft RPG.
5 Following public consultation on the proposed changes, the Secretary of State issues the final RPG.

The process then continues via monitoring and review. The process began in all regions between 1991 and 1996 (Counsell, 2001) and in most regions the final RPG was approved between 2000 and 2002. Revised guidance, notably with regard to more rigid timetables for each stage and broader policy concerns, was published in 2000 in PPG11.

Table 1.5   Main concerns and objectives of RPG in England and Wales

Region	Main concerns
Northern	Urban economic regeneration. Environmental enhancement
Yorkshire/Humberside	Sustainable development
North West	Sustainable economic development. Competitiveness, prosperity, quality of life
West Midlands	Sustainable development and growth
East Midlands	Ensuring that growth and development is sustainable
East Anglia	Achieve environmentally sustainable growth
South East	Need to meet environmental objectives; economic performance. Importance to rest of UK
South West	Sustainable development
Wales	Sustainable development. Regulate development

Source: Roberts (1996)

Table 1.5 shows that the main concerns and objectives of the first round of RPGs in the mid-1990s focused around sustainable development, but with growth and regeneration surfacing as sub-issues. The main areas of dispute have been the number of houses to be built where they are to be built, and the percentage of housing land that will be allocated on so-called greenfield sites, i.e. land not previously developed. Once agreed, these decisions have to be followed by Structure and Local Plans. The situation in 2002 for England is shown in

Table 1.6 Key regional planning documents in England (outside London), 2002

	Regional planning guidance	Regional economic strategy	Regional sustainability framework
Prepared by	Draft prepared by local authorities in the region, acting collectively. Regional planning conferences/new regional chambers (local authority members and at least 30% representatives of business and the voluntary sector)	Regional Development Agency. Eight to fifteen members appointed by Ministers, chair and majority of members with business experience, four members from local government, funded by the government office for the region	Government office or regional round table on sustainable development
Form of scrutiny	Public examination, arranged by government office, which then consults on any proposed modifications	None	Public examination. No formal process of assessment or approval
Adopted by	Secretary of State	Regional Development Agency	Regional Chamber
Time scale	Fifteen to twenty years	Five to ten years	None specified?
Guidance on preparation issued by	DETR in PG11, revised 2000. From 2001, DTLR	DETR till June 2001, since then DTLR	DETR 2000
Consultation procedures	Formal consultation stages, public conference on draft, draft RPG submitted to Secretary of State, proposed changes to draft RPG. Community participation is encouraged, especially of Local Agenda 21 groups	Regional Development Agency must have regard to views of, and be willing to give an account of its activities to, Regional Chamber	None specified
Date produced	New style RPG began in 1999	April–October 1999	2000 and 2001
Date revision due	Continual review and annual monitoring of implementation impacts, where possible	Statutory requirement to keep under review	Discretionary
Statutory basis	Non-statutory	Statutory	Non-statutory

Source: Royal Commission (2002)

Table 1.6. As Chapter 5 will show, regional plans became very important from 2004 onwards and by 2010 Regional Planning Assemblies may well have become the dominant power in planning.

In Wales and Scotland regional planning guidance has been replaced by national strategic guidance prepared by their elected assemblies and parliaments set up in the late 1990s as part of devolution. Wales had a form of strategic guidance in the mid-1990s, as shown in Figure 1.8.

In all three countries the new system has made the democratic deficit between strategic and local planning even worse, since most people are unaware of the implications of the Regional or National Guidance, which receives very little media attention. People only become aware of the issues too late in the process when a new development is proposed on their doorstep, but protest is largely useless, since the decision in principle has already been made.

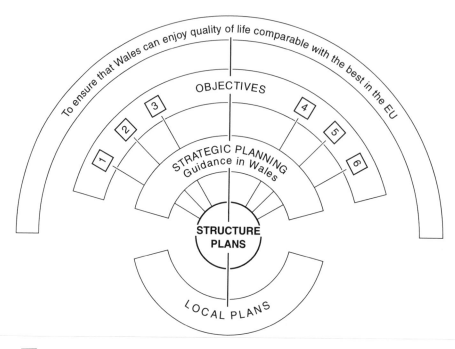

1 Developing land use planning policies which reflect the principle of sustainable development

2 Protecting and enhancing the natural and built environment

3 Recognising the distinctive language and culture

4 Improving the economic health of the principality

5 Improving access to housing

6 Securing public and private investment in transport and telecommunications infrastructure

Figure 1.8 The hierarchy of planning policy guidance in Wales in the mid-1990s. *Source: Jarvis (1996)*

Summary of changes since 1947

There have been four main periods:

1 *1947 to the early 1970s.* Development Plans based on a detailed map, setting out exactly where development will take place. They were called Development Plans because literally they showed where Local Authorities would build things in a socialist world.

2   *The 1970s.* Gradual introduction of Structure and Local Plans based on broad regional frameworks. Major reform of local government.
3   *The 1980s.* Introduction of Unitary Development Plans in the main metropolitan areas, elimination of regional land use plans.
4   *The 1990s to date.* Reintroduction of regional plans with a gradual increase in power till by 2002 they set the agenda for sub-regional and local planning. Muddled reform of local government in England but more rational reform in Wales and Scotland.

The situation in 2002 is shown in Box 1.7.

Box 1.7   Development Plans in 2002

Development Plans have two core components – a Strategic Plan and a more detailed Local Plan – but with the following detailed arrangements:

1   In areas of England with a two-tier structure, counties prepare a Structure Plan and districts a Local Plan which must conform with the Structure Plan.
2   Unitary authorities in Scotland produce a Structure Plan and a Local Plan.
3   Some authorities in England and Scotland have been required by central government to produce a joint Structure Plan with one or more neighbouring authorities.
4   Unitary authorities in England and local authorities in Wales produce a Unitary Development Plan in which Part 1 is broadly equivalent to a Structure Plan and Part 2 to a Local Plan.
5   In National Parks in England and Wales the park authorities produce the Development Plan as well as a management plan.
6   In England, county councils or unitary authorities prepare Waste Plans and Minerals Plans. The Scottish Executive has proposed that Waste Plans should be introduced into Scotland.

All the components of a Development Plan are subjected to extensive consultation. If objections to a Local Plan are pressed there is a public inquiry conducted by an inspector (England and Wales) or an inquiry reporter (Scotland). A Structure Plan is considered at an Examination in Public conducted by a panel chaired by someone appointed by the government. The local authority considers whether to modify the plan in the light of the report by the inspector, reporter or panel. Ministers can intervene at any stage and direct changes prior to its adoption. In Scotland Structure Plans require the Minister's approval.

Source: Royal Commission (2002)

The broad principle of having a plan has remained intact, but the details have changed considerably. In particular, plans became guides for the control of development and thus depend for their implementation on the actions of developers and business people. Their effectiveness can thus be judged only by a study of development control as set out in Chapters 3 and 4. It is thus now appropriate to study how the development control system has evolved.

THE DEVELOPMENT CONTROL SYSTEM ——————————

Four key points about development control

At the outset it is important to make four key points about development control:

1   It has stood the test of time and has remained essentially unaltered except for detailed changes since its inception in 1947. This must say something about its quality. It also means that this section does not take an overtly historical approach except where there have been changes. However, an evolutionary approach is taken at the core of this section, which describes the sequential stages in making a development control decision.
2   It is the 'sharp end'/'coal face' of planning. It does not have the glamour of plan making and does not tend to attract theoretical 'high-flyers', but for practical planners it does have the attraction of being able to influence individual decisions.
3   It is where the impact of planning is felt and can be measured in terms of what has and what hasn't been allowed, especially with regard to the effect of the system on the geography of Britain.
4   But like birth control its effects are hard to assess, because what hasn't happened can't be known. This can be called the 'deterrent effect' in that planning applications are not made if there isn't a reasonable chance of success. In addition there is the 'displacement' effect which has affected behaviour by displacing planning applications elsewhere. The 'deterrent' and 'displacement' effects thus make measuring the effects of development control a methodological nightmare.

Nonetheless, we need only look at the sprawling suburbs of the 1930s and individual houses built along main roads in the countryside before the introduction of development control in 1947 to see what might have happened. Alternatively, we can look at countries that have laxer systems of control, like Spain or the United States, to see how different our environment might have been, in terms of greater urban sprawl and enormous roadside advertisements, for example.

The power of development control lies in its ability to permit, refuse or impose conditions on planning applications for the development of land. But what is 'development'?

Definition of development

The Planning Acts define development as 'The carrying out of building, engineering, mining or other operations, in, on or over the land' or 'the making of any material change in the use of any building or land'. Thus in theory every change of land use comes under development control. This would be nonsensical, and so the definition of development is deliberately vague so that the government can alter the exact definitions of the type of development that need planning permission from time to time in either the General Development Order or the Use Classes Order.

1   *The General Permitted Development Order (GPDO)(1995)*. The GPDO sets the limit *outside* which planning permission is needed in each category of development. For example, minor changes in the size of a house are *below* the limit and do not need planning

permission. In addition the GPDO also provides exemptions from planning permission in each category For example, most developments on farms used to be exempt till recently. The GPDO is changed frequently to accommodate changes in society, for example the introduction of mobile phone masts in the 1990s. The GPDO is published as a Statutory Instrument, which is approved by Parliament. They can be found in public and law libraries and are catalogued by ascending number in each year with an S.I. prefix.

2   *The Town and Country Planning (Uses Classes) Order (UCO)(1987).* The UCO divides changes of use into categories; changes within categories are normally exempt, unless big changes are made, when they would probably come under the GPDO. In common with the GPDO minor changes of use may be exempt. Also in common with the GPDO the UCO is regularly updated to accommodate change.

Because both the GPDO and the UCO are legal documents with exact wording they are difficult to read and so they are usually accompanied by circulars from the government. These put the text into plain English and also provide guidance on how the changes are to be interpreted.

The development control process (from application to final decision)

The key stages in the development control/planning application process are shown in Figure 1.9. Each stage is now considered in turn. The application (submission) is made to the local planning authority (LPA). Most applications are thus made to district councils or in the National Parks to the National Park Authority. Mineral applications or those involved with waste disposal go to the counties. Since 1981 a fee has been payable for what is now regarded as a service, but it does not normally cover the full cost. The fee is related to the size of the application. Two types of application can be made: outline or detailed.

An *outline* application deals only with the broad principles of the proposed development, e.g. whether houses can be built or not. The advantage for applicants is that they can test the waters with the minimum of outlay on either architects' drawings and even purchasing the site, since an applicant does not have to own land to apply for planning permission. For the planner the stakes are low, so that genuine negotiation can take place. The disadvantage for the planner is that once the principle of permission is conceded only details can subsequently be negotiated. Detailed applications can be made at the outset or after an outline application has been approved.

Development control is also a public process, and applications are available or advertised for public comment. In addition, the owner of the land involved must be informed, since otherwise vast profits could be made by the back door. For example, farmland may be worth £5,000 per hectare, but building land may be worth at least £250,000 per hectare. Thus if landowners were not told that planning permission had been obtained they might sell it unwittingly at a large loss, since applicants do not have to own the land to apply for permission to develop it. Even so there have been a few unfortunate cases where unscrupulous lawyers have duped over-trusting or infirm clients into selling land as if it did not have planning permission. Notices on lamp posts and advertisements in local newspapers also give publicity. There is no statutory obligation to inform neighbours, although it has often been advocated, but in fact most planning authorities inform neighbours as good practice. Instead it is argued

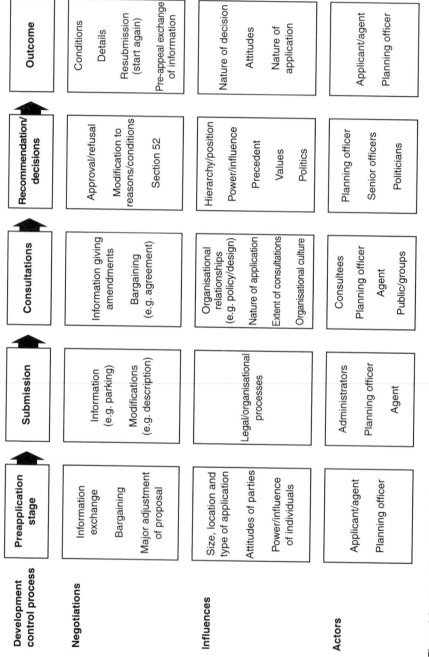

Figure 1.9 Negotiations in the development control process. *Source:* Claydon (1996)

that planning is in the public interest and is not meant to be a neighbour dispute resolution service.

The decision-making process on a planning application is influenced by three generic factors over and above the facts in each case:

1   *Government advice on how to interpret the legislation.* For example, in 2000 advice from the government in a Planning Policy Guidance Note on Housing PPG3 advised that new housing should be restricted to existing towns and cities, with 60% of new housing on brownfield land by 2008. There should be a sequential approach to new housing in which inner city or derelict land sites are examined first, then suburban infill, then edge of town sites and finally rural sites. The PPG also advised that densities should be between 30 and 50 houses per hectare. In addition to formal PPG advice, there are also circulars, speeches, White Papers, etc. Before the introduction of PPGs in the late 1980s advice was given in circulars and continues to be given when the subject matter is not broad enough to be the subject of a PPG. For more details see Boxes 1.5 and 1.8.

Box 1.8   Example of advice on development control

DOE Circular 22/80, *Development Control Policy and Practice*:

> The Government's concern for positive attitudes and efficiency in development control does not mean that their commitment to conservation is in any way weakened: in particular they remain committed to the need to conserve and improve the countryside, natural habitats and areas of architectural, natural, historical or scientific interest and listed buildings. There is no change in the policies on national parks, areas of outstanding natural beauty or conservation areas.
> The Government continue to attach great importance to the use of green belts to contain the sprawl of built-up areas and to safeguard the neighbouring countryside from encroachment and there must continue to be a general presumption against any development within them. Nor will the Government allow more than the essential minimum of agricultural land to be diverted to development, nor land of a higher agricultural quality to be taken where land of a lower quality could reasonably be used instead.

In 1987 Circular 16/87 modified this advice to 'the continuing need to protect the countryside for its own sake rather than primarily for the productive value of the land'.

In 1992 PPG3 (Housing) and PPG7 (*The Countryside and the Rural Economy*) made further modifications:

1   PPG3 put increased emphasis on reusing urban land as a means of relieving pressure on the countryside.
2   But PPG7 argued that, though buildings in the open countryside should be strictly controlled, elsewhere developments which benefited the rural economy were OK.

---

2   *Planning case law.* As Chapter 2 will outline, British law is based on precedent and Acts of Parliament use general terms and at times vague language on purpose, for example words like 'amenity', the 'public interest' and 'material considerations'. This allows exact meanings and definitions to be set out in GPDOs and UCOs, for further definition by the legal definition and maximum flexibility for local planners. Thus there is a rich literature of legal decisions about how to operate development control. These can be

found every week in *Planning* magazine and every month in the *Journal of Planning and Environment Law*. In the longer term these decisions are regularly updated in expensive loose-leaf binder-type volumes with names like *The Encyclopaedia of Planning Law* which can be found in legal libraries and in all LPAs. Increasingly, databases on CD-ROM or the Internet are also being made available but are expensive.

3   *Planning case lore, received culture from colleagues and training courses*. Planners will often phone around colleagues to get informal advice. Every planning office has its own 'culture' and stories told over a pint of beer or on the conference circuit. It is unwise though to assume that any two planners will come up with the same recommendation.

Turning to how decisions are made in each case on a specific basis, one key factor derives from the Planning and Compensation Act 1991. Section 26 of the Act introduced a *presumption in favour of development proposals which are in accordance with the development plan unless material considerations dictate otherwise*. This ushered in the so-called 'plan-led' system as from 1992. The change may have been more apparent than real because before the 1992 Act advice had referred to Development Plans and both Acts and circulars had emphasised two key points:

1   Decisions must have regard to the provisions of the Development Plan.
2   The onus was on LPAs to show that applications should not be allowed, thus there was an expectation that permission would be granted.

The key change in 1992 was thus from 'having regard to the provisions of the plan' to one of having a 'presumption in favour of proposals in accordance with the plan'. In either case decisions contrary to the plan can be and are made. Indeed, this is the hallmark of the British system, compared with zoning systems practised elsewhere, in that in Britain each application is treated on its own merits as a one-off. In this regard the key phrase in the 'plan-led' system are the 'material considerations'.

Thus we now need to consider what 'material considerations' are and how they are weighted against the need to take the plan into account, and to presuppose that permission should be given. In this process three factors may be important: checklists, the public interest and the possible need for an Environmental Assessment.

1   Checklists may be used either formally or intuitively. At a basic level, checklists will confirm that everybody who should be consulted has been and that every procedure that should have been done has been. More important, in formulating a recommendation four or more things may be checked by the planning officer:
    (a) Does the application conform to the plan?
    (b) What effect will the proposal have on road safety, amenity and public services?
    (c) Is there a conflict with public proposals to use the same land?
    (d) Is there a need to protect natural resources in the area, e.g. sand and gravel?
    (e) Other relevant considerations, for example the effect of a new out-of-town supermarket on the city centre.

2   The checklist will then be aggregated into a view of what is in the 'public interest'. Planners would not normally consider as material, for example, the personal circumstances involved, however heart rending they may be, for example, the desire to build a 'granny flat' for an ailing relative. Likewise, reductions in property values or sales

lost to a new supermarket are not material considerations. Finally, the views of neighbours, while useful in formulating a decision, should never be the crucial part of the decision. The final decision should rest on the question 'Will the proposal lead to an unacceptable change in the character of the area?' in the context of all the other issues, like PPG advice, outlined earlier.

3   An aid to this assessment may be provided by an informal or formal 'Environmental Impact Assessment'. These originally derived from experience in the United States, where they were invented to weigh up the pros and cons of major developments like the proposal to build a major oil pipeline in hitherto unspoilt Alaska in the late 1960s. They were introduced into European legislation in the 1980s and are a rare example of European legislation impinging on British planning legislation. Indeed, the British government resisted the proposals, arguing that the processes outlined above amounted to an implicit environmental assessment.

The legislation divides developments into two types, first those where an assessment is mandatory (Annex 1) and second, those where an assessment may be made (Annex 2). The applicant makes the assessment. This involves a balance-sheet-style approach where the costs and benefits of the proposals are weighed against each other. The 'bottom line' may not, however, be as precise as the numbers imply, since the assessment will have included lots of problematic assumptions, for example the value of wildlife or historic churches. Nonetheless, as an advanced form of checklist they do offer decision makers information on which to make a more informed decision. Jones and Bull (1997) report that there were 1,827 statements between 1988 and 1994, averaging 260 a year, but that the number had risen from 243 in 1990 to 347 in 1994. Annex 1 assessments account for 10% of statements and waste disposal assessments are the most common.

In a major review of the accuracy of the assessments Wood *et al.* (2000) found that about 80% of predicted environmental impacts had been accurate, with very few unpredicted impacts. They concluded that the system was working but needed to be improved via a number of changes, notably more explicit predictions.

Now that we have examined how planning officers come to make a recommendation it is now time to examine the types of recommendation they can make in the context of the key stages of the process shown in Box 1.9.

We have already considered the consultation and preliminary formulation of the recommendation shown in stages 1 and 2 of Box 1.9. Alongside these two stages the third stage, dialogue or bargaining with the developer, has become increasingly important in recent years, but we will consider it in more detail below under the generic title of 'conditional permissions'.

Once the planning officer has completed stages 1–3, which in practice are carried out concurrently, the next step is to decide who should make the decision. There is plenty of scope for delegating decision making to the planning officer according to local circumstances. For example, if the application is small and non-controversial, often a so-called 'householder' application (a house extension, for example) the decision can be made by the planning officer under delegated powers. More complex and larger applications go to the Planning Committee, which will normally meet every month. In most cases the officer's recommendation is accepted. Very large or controversial applications will go to the full council. In all cases the reasons for the decision must be set out in a decision letter which is the legal document.

Box 1.9 Key stages in the development control process

1 *Consultation.* Formal/informal, with all those involved.
2 *Consideration.* Public interest, quality of work, etc.
3 *Dialogue with developer.* Planning 'gain' that may be required of the developer, e.g. a new leisure centre or access road.
4 *Recommendation.* Made by the officer, and the decision may be delegated to officers, or may go to the Planning Committee, or infrequently to the full council.
5 *Decision.* Should be made within two months.

The decision can take three forms:

1 Unconditional approval (within five years).
2 Conditional permission:
  (a) 1971–91. Bargaining under Section 52 of the 1971 Act.
  (b) 1991 onwards. Conditions accompanied by planning obligations under Planning and Compensation Act 1991 and Circular 16/91 + 11/95.
3 Refusal.

As Box 1.9 shows, three types of decision can be made: unconditional permissions; conditional permissions and refusals. *Unconditional* permissions are not very common and indeed are not, strictly speaking, unconditional, since they have a time limit of normally five years.

The most common type of decision is *conditional permission*. Conditions according to DOE Circular 11/95 should be necessary, relevant, enforceable, precise and reasonable. These terms are as before deliberately vague to allow local discretion, but they do of course open up the potential for appeal as discussed later. As Table 1.7 shows, conditions vary by the type of area, with visual conditions being more important in designated areas like Areas of Outstanding Natural Beauty (AONB). The table also shows that several conditions may be imposed, with on average around two types of condition per permission.

Table 1.7 Conditions on permissions used in rural East Sussex

	Frequency of use			
	Sussex Downs AONB (36 applications)		Low Weald (284 applications)	
Conditions by category	No.	%	No.	%
Visual amenity	31	50.0	224	39.3
Future regulation of the site	16	25.8	138	24.2
Highway safety	11	17.7	130	22.8
Residential amenity	4	6.5	51	8.9
Site regulation	0	0	27	4.7
Total	62	100.0	570	100.0

Source: Anderson (1981)

However, conditions can fail and have bad side effects. For example, developers may not follow the conditions, and planning departments do not have enough resources to go out and check buildings when they are being constructed. Once built it can be difficult to get builders to implement any conditions that may not have been followed. Other conditions while well meaning can have bad side effects. For instance, in high-value landscape areas, conditions will often be imposed that buildings should be built of local materials. These will usually be more expensive, and the resulting high cost may well force local people out of the housing market. To counteract this planners may impose conditions that a certain percentage of a development may be for 'affordable' housing for people with a low income or for 'local' people.

Conditions can also play a creative role by turning a poor application into a good one. Planners can also use the concept of planning 'gain' and use 'obligations' to extract public benefits from an application. For example, planners may try to get developers to provide some sort of social housing, a community facility, a new road or a recreational facility. Research by Campbell *et al.* (2000) has shown that less than 2% of all applications involve an obligation, but that 18% of major developments attract an obligation. Major residential developments attracted the highest rate of obligations at 25%. Negotiations over these issues may often take place informally between developers and planning officers to test the water. While this is laudable, the process has been criticised.

First, some people say that planning 'gain' is virtually the sale of planning permission. Second, it encourages a different attitude to development, which may put financial interests over objective planning considerations. Third, because the process may have to be secret early on because of the commercial sensitivity of say a new supermarket, planners may be open to suspicions of graft or corruption. Fortunately, cases of planning corruption are rare, but anything which compromises planning's reputation is a cause of concern. Overall, Campbell *et al.* (2000) are concerned that the use of obligations has begun to influence decision making, with short-term financial and other gains tending to override long-term concerns like environmental quality. They conclude that these trends could fundamentally challenge our conception of planning from a public service towards one geared to providing infrastructure and financial gains.

The third type of decision is a *refusal*. A refusal is the least common because of three factors. First, there is the 'deterrent' effect pointed out at the outset of the chapter. Many applications simply aren't made and others are not made following informal advice from planners. The great imponderable is how much has been prevented. There is no way of knowing. Second, if a refusal is upheld and the council is deemed to have acted too harshly, applicants can apply for costs to be awarded against the LPA, which is potentially very expensive. Third, councillors may see a development bringing lucrative tax increases, more jobs and so an electoral advantage from giving a planning permission. In addition many councillors are in business or favour the group of interests they come from. Thus one or two councillors can push for permission, while the rest of the committee sit idly by, as demonstrated by Gilg and Kelly (1997a).

If a refusal is made, the reasons for refusal have to be clearly spelt out so that the refusal can be defended at any subsequent appeal. The soundest reasons are those based on policy but as Table 1.8 shows other reasons may be used, thus reinforcing our earlier point that each application is treated on its merits, and is a 'one-off' decision. As with conditions there are differences between the two areas, except for non-policy reasons which are pretty similar. There are also more reasons per application, with between three and four reasons per application.

Table 1.8   Reasons for refusal of planning permission used in rural East Sussex

Reasons by category	Sussex Downs AONB (29 applications)		Low Weald (322 applications)	
	No.	%	No.	%
Policy Reasons				
Settlement policy	27	28.4	251	36.9
AONB	14	14.7	0	0
Countryside buildings	5	5.3	49	7.2
Countryside conservation	3	3.2	16	2.3
Town policy	0	0	17	2.5
Reasons other than policy				
Site regulation	17	17.9	100	14.7
Visual amenity	16	16.8	113	16.6
Highway safety	7	7.4	101	14.9
Building conservation	5	5.4	10	1.5
Future regulation of site	1	1	23	3.4
Total	95	100.0	680	100.0

Source: Anderson (1981)

Refusal though not common brings us to an extra possible stage, the involvement of the government via the *appeal system*. An appeal can be made by the applicant on several grounds. First, that a decision has not been made in two months, though appealing is only going to make a decision take longer. Second, an appeal that the conditions are too onerous. Third, an appeal against refusal.

Appeals are dealt with by the Planning Inspectorate, an executive agency of the government. Nominally all its recommendations are dealt with by the Secretary of State, but in practice most decisions are delegated to the inspectorate. Important or controversial decisions depending on increasing importance or controversiality are made by civil servants, Junior Ministers, the Secretary of State or even the Cabinet. The decision is final unless maladministration or a legal mistake can be found. Decisions are based on evidence presented at a public inquiry, in a private hearing or by written evidence.

1   A public inquiry is the most formal of the options, in which a quasi-judicial atmosphere prevails, with formal cross-examination by legal advocates on behalf of the main parties and evidence from experts, which may take several days, months or even years.
2   A private hearing is an informal process in which the inspector listens to those involved in a small room, and professional help and consultants may be used.
3   Written evidence is the simplest type of appeal, in which the participants provide reports on their respective points of view. Written evidence also forms a major part of the other two processes.

Most cases are simple and dealt with by written representations (normally 80%) or an informal hearing (normally 12%). Only around 8% are complex or contentious enough for a public inquiry and only 2% are decided at the ministerial level. The number of appeals is also strongly related to the economy, notably the state of house prices, and in a house price boom the number of appeals rises, as in the late 1980s to reach a peak of 32,000, but in the dramatic slump of the early 1990s appeals fell substantially to only 18,000. Appeals also rise if the government is felt to be sympathetic, as in the mid–1980s when a number of successful appeals encouraged more appeals.

Whichever process is used, the planning inspector writes a report from which a decision is made, as shown in Boxes 1.10 and 1.11. In common with decisions at the local authority level, the decision can be unconditional permission, can attach conditions or can refuse the appeal. Reasons must be given for each decision.

Box 1.10 Case study of an appeal decision: golf course

The developers wished to construct two championship golf courses and associated facilities, including a hotel and leisure complex, but had been refused permission. The appeal site was a common with heathland vegetation in an Area of Outstanding Natural Beauty used for rambling by local people and was close to a seaside resort in decline. At the appeal their QC argued:

1   Golf courses would increase access. Golfers would be new users, and ramblers could use paths through the course.
2   There was an estimated need for one new golf course a week, especially in a declining tourist area, to provide alternatives to beach holidays and to reduced farm jobs.
3   Although the area was 'common land' such land is privately owned, and use by the public could be rescinded at any time by the owners, especially as such use costs the owners money in repairing damage, putting up signs, etc.

The protestors, who were represented by formal pressure groups like the Council for the Protection of Rural England and an *ad hoc* pressure group formed to fight the proposal, appointed a QC, who made the following points, which by and large endorsed the case made by the LPA planning officer:

1   Wild areas in southern England are rapidly disappearing, and thus it was extremely important to stop decline.
2   Fifteen thousand people had been recorded using the area over a one-month period.
3   Over 30,000 people had signed a petition against the proposal.

Six months after the two-week-long inquiry the following decision was issued:

> The appeal site lies wholly in an Area of Outstanding Natural Beauty. Such areas must be preserved as far as possible against intrusions, which are likely to detract from their character. The wild and natural landscape of the two commons conveys an impression of remoteness and peace. The two golf courses with their ancillary buildings and car parks would have had a discordant impact on the natural beauty of the heathland. Although some parts of the proposed courses would have been open to the public and paths could have been used for riding and walking, there would have been an unacceptable loss of freedom for visitors wanting to wander at will. The need for golf does not outweigh these disadvantages.

Key point. The decision was made via a balancing process, weighing golf against the broader public interest.
Postscript. The developers re-applied for permission but on farmland adjacent to the heathland. This time permission was given on the grounds of job creation for the declining farm and tourism industries, and that the golf course would have a richer habitat than farmland.

Source: based on the local press and the *Journal of Environment and Planning Law*

Box 1.11 Case study of a planning appeal: oil refinery

An oil company wanted to build an oil refinery on Canvey Island in the Thames estuary. This would have been detrimental to the residents in terms of not only their quality of life but also physical danger in the event of an accident, since escape from the island would be restricted.

Because of the national nature of the decision the government 'called in' the application for a decision. In such cases it is felt that local government is not the appropriate forum for a decision.

After a lengthy public inquiry, the decision was made by the Cabinet, to the dissatisfaction of the Secretary of State, who wrote in his memoirs (Crossman, 1975):

> Normally I would have decided that planning application without consultation in favour of the unfortunate house owners. But the national economic interest was so big that I allowed the Dame (Evelyn Sharp) to bring the issue to Cabinet, where I was duly overridden by the collective decision.

Key point. Balancing can be not only between issues but also between organisations. In this case the financial Ministries won out over the environment Ministries and the economy won out over conservation.

Another type of public inquiry is occasioned by 'call-in' applications when the government decides that the issues are too complex for a local decision or there are national issues at stake. This process has been queried in that the government is thought to be acting as judge and jury and is at present subject to change dependent on appeals by pressure groups and others under the European Human Rights Act. Typical 'call-in' applications have involved airports, seaports, mining and major roads. Really big cases like the Channel tunnel and the rail link to the tunnel now under construction are decided by Acts of Parliament. Other cases involve applications where the proposal is a substantial departure from the Development Plan or many applications are being made and only one will be allowed. For example, most towns have had to decide on the siting of major out-of-town retail and commercial developments. In some cases up to 13 different proposals have been put forward. The 'call-in' procedure is thus a very sensible option in such cases.

Variations over place and time and special controls and measures

The strength of planning control varies by the type of development and over space and time. The main variations are as follows: national protected areas; the agricultural exemption to the GPDO; Enterprise Areas; Simplified Planning Zones; and special controls like Article 4 Direction Orders, listed buildings, Conservation Areas and Tree Preservation Orders. In addition there are special measures to deal with breaches of control and special rights for the public. More detail on how these policies provide spatially different impacts is provided in Chapter 4.

National protected areas

Around 30% of Britain is covered by some sort of protected area policy, most notably Green Belts, National Parks and Areas of Outstanding Natural Beauty. In these areas there is a presumption that planning permission will be more difficult to get or will be subject to stricter conditions. The areas do not mean that permission will never be given, but that a special case has to be made.

The agricultural exemption to the GPDO

Because of the alliance between town planning and agriculture caused by the Second World War, farm buildings were given a major exemption from planning control so that agriculture would not be fettered in any way. In addition new dwellings for farm workers have been given a major 'exception' in local planning policies. This has allowed major landscape changes, even in protected landscapes. However, in the 1990s the policies were made less favourable to farmers. First, in the early 1990s in the National Parks farmers had to notify the planners if they wished to construct a new building and planners had the opportunity to impose planning controls if they so desired. Second, this power was later extended to all areas. Third, the GPDO exemption was much reduced. The 'farm workers' exception is still, however, open to abuse and Gilg and Kelly (1996a) have shown how farmers can get permission by building a bogus case for a farm worker's house usually by dividing a farm into two with only one half having any houses. There is thus an apparently cast iron case for building a house on the farm without a house. Subsequently, the new house is sold off to someone else. Often the house is also substantially larger than needed for the farm. Councillors have been happy to turn a blind eye to this practice in many areas, as shown in Chapter 4.

Temporal differences in the GPDO

In the 1980s the Tories increased or simplified the limits in certain cases in order to (as they put it) 'lift the burden' of planning from free enterprise. For example, in Statutory Instruments 245/1981 and 246/1981 they increased the percentage by which houses and factories could be extended without needing planning permission from normally 10% to normally 15% but in some cases to 20% or 25%. The percentage can be used only once. This has had obvious effects in the suburbs, with a rash of conservatories and other extensions no longer needing permission. Whether this has improved the suburban environment is a matter of opinion. The changes did not, however, extend to National Parks, Areas of Outstanding Natural Beauty or Conservation Areas.

Enterprise Zones, Urban Development Corporations and Simplified Planning Zones

Enterprise Zones were introduced by the Local Government, Planning and Land Act 1980. In the zones planning controls were virtually abandoned in favour of simple zoning rules. The idea was that inner cities and derelict industrial or port areas could be revived only if controls were effectively scrapped in a sort of 'Hong Kong' effect. Thus zones were created, for example, in the London docks and on a former steelworks in Corby, Northamptonshire. A variation on Enterprise Zones has been provided by Simplified Planning Zones (SPZs) under the Housing and Planning Act 1986. In these areas local planning authorities can draw up a list of developments that are exempt from planning control, or require less rigorous control. However, the measures have not been widely used, largely because local government has been in the hands of the Labour and Liberal Democrat parties, which have been ideologically opposed to giving up power. There has also been a lack of entrepreneurial demand for the designation of SPZs, since axiomatically these areas are not economically dynamic. In parallel with Enterprise Zones, Urban Development Corporations and City Action Teams were set up in order to regenerate run-down inner cities

The immediate reaction according to Norcliffe and Hoare (1982) and Purton and Douglas (1982) was that the zones would merely siphon off existing developments or that they would re-create the horrors of the Victorian city. At first this seemed to be the case, with major newspapers abandoning Fleet Street and moving to the former dock area of Wapping. But by the 1990s some of the areas had created substantial numbers of jobs, notably the old dockland area in London known as Canary Wharf. Successor schemes like the Cardiff Bay development have also been successful economically and socially, albeit controversial environmentally in that they necessitated the drainage of large wetland areas of considerable wildlife interest.

Article 4 Direction Orders

In contrast to Enterprise Zones, Article 4 Direction Orders bring operations that are normally exempt from planning control under the GPDO into control. The following provide three examples of where such controls have been used. First, Royal Crescent in Bath, where the harmony and symmetry of the classical Georgian crescent is partly created by most of the doors being painted black. However, one owner in the 1970s painted the door yellow, and so to re-create the harmony of the crescent the LPA invoked an Article 4 order rescinding the normal freedom to paint doors any colour. (Note that since this case in the 1980s, Listed Building Control (see next section) has brought the painting of listed buildings under control, so that an owner seeking a colour change must seek planning permission.)

Second, some farmers were making money from dividing up fields and selling them as small 'leisure plots'. On these plots owners could erect small buildings under the GPDO. The resulting 'allotment'-style landscape was unattractive and so where it was happening LPAs invoked Article 4 orders to bring such development under control. Third, altering houses in response as much to marketing and fashion as to any practical benefits, exemplified by uPVC windows, 'stone cladding' and the removal of architectural features, has become a major feature of inner-city terrace houses as owners seek to make them different or more attractive. Most planners disagree and so they have often brought 'stone cladding' under control via an Article 4 order. This is a matter of taste, however, and may involve unusual developments. For example, a house in Oxford featured a life-size plastic shark diving into the roof. After much controversy the shark was removed under an Article 4 order.

Listed buildings and Conservation Areas

Although legislation to preserve important (listed) buildings began under the Town and Country Planning Acts the legislation for them had become so complex that they were given a separate Act of Parliament in 1990: the Planning (Listed Buildings and Conservation Areas) Act 1990. This consolidated and modified existing legislation. Under the Act the relevant Minister has a duty to compile a list of buildings of special architectural or historical interest, hence the title 'listed buildings'. Around half a million buildings have been listed. They are not necessarily beautiful or old: for example, some 1930s Tube stations and cinemas and Centre Point in London are listed. They are ranked into three grades, with some 8,000 in the top Grade 1 (Larkham and Jones, 1993).

If a listed building is to be altered or demolished listed building consent is needed. This can include the colour of exterior walls. Modern materials are not by and large allowed, notably plastic or metal double glazing. The power to prevent demolition or alteration is thus a key power over listed buildings. If a non-listed building is threatened a Building Preservation Notice gives six months' grace while a decision over listing is made, and also time to record or remove valuable features. A counter-provision relating to unsafe buildings allows builders to arrive at 4.00 on a Sunday morning to find that the building is suddenly dangerous, or that it has caught fire, or that a lorry has driven into it, etc. The damage is done, the fine may be small, the proof difficult, the extra profit vast. Long-term neglect is a more insidious enemy. The key to preservation is after-use and positive aid, but cash aid is limited by total and is also means-tested. Another problem is that VAT is chargeable, thus increasing repair and maintenance costs, which are anyway more expensive than for normal buildings by a further 17.5%.

When there is a collection of listed buildings a Conservation Area can be designated which provides a more cohesive form of protection. Between 200 and 500 Conservation Areas were designated a year throughout the 1970s and 1980s. The idea is to preserve the setting of the most important buildings, because every building in a Conservation Area enjoys heightened protection and is subject to more stringent control as a result.

Tree Preservation Orders

These prohibit the cutting down, lopping or wilful destruction of a tree, groups of trees or woodland. If felling is allowed there is usually a replanting condition. However, trees and woodland can suffer as much from neglect. For example, if hedges or fences become derelict, livestock can enter and graze young saplings, thus preventing regeneration. Developers may also destroy trees because the fines are not severe enough or may 'accidentally' allow trees to die by letting waste products seep into the ground near by or by an 'accidental' collision with machinery. Nonetheless, Tree Preservation Orders provide a sanction and can be invoked almost instantaneously if trees are threatened with felling.

In spite of the rhetoric of the Tories between 1979 and 1997, controls have in many cases been tightened, notably in the Planning and Compensation Act 1991. The main controls are concerned with preventing breaches of planning control, notably enforcement controls. The controls are balanced by a set of individual rights.

Enforcement controls

The first control is an Enforcement Notice, which is served if a development has taken place without planning permission or has been built contrary to conditions. If an early stop to an activity is needed a 'stop notice' can also be served. Planning Contravention Notices and Breach of Condition Notices are related legal remedies. These notices are very necessary since there have been a number of cases where breaches have occurred. For example, there was the case of a Cornish farmer who, after being refused permission to build a retirement bungalow, constructed a stable for a pony. The stable looked remarkably like a bungalow except for the wall-to-wall straw! The LPA served an Enforcement Notice but the farmer appealed. The appeal succeeded on humanitarian grounds but the decision letter made it clear that no precedent had been set.

But these powers can be unpopular and are used only in straightforward cases or when there has been a significant breach of planning law. There are three main reasons. First, the LPA may not want to stimulate controversy and appear overbearing and unreasonable. Second, enforcement can be unpleasant and in extreme cases can lead to death. For example, in the 1990s a planner serving an Enforcement Notice on the owner of an illegally built dwelling was shot dead beside the bulldozers that had been called in to demolish the building. Third, enforcement, especially demolishing property, is poor publicity for planning but makes good 'copy' for readers of newspapers that proclaim freedom from state control like the *Daily Express* and the *Daily Mail*. Ironically, it is the often obsessively neat and tidy suburbia inhabited by such readers that planning protects.

Individual rights

There are four main rights for individuals: a 'Purchase Notice', a 'Blight Notice', an application for a 'Certificate of Lawfulness of Existing Use or Development' and an application for a 'Certificate of Lawfulness of Proposed Use or Development'. A Purchase Notice may be served by an individual on an LPA when they feel that the use of their property has been so severely impaired by planning decisions that they can longer use it beneficially. If successful the Purchase Notice forces the LPA to purchase the property. An obvious problem then arises as to how the property should be valued, before or after it was affected? If there is a dispute a third-party arbitrator may be brought in. A particular type of property loss is caused by 'planning blight' when uncertainty over the future of an area brought on by planning delays makes a property unsaleable. In that case an individual can serve a Blight Notice on the LPA, which has the effect of leading to a Purchase Notice if successful.

If an individual has constructed a building or changed the use of a building or an area they can retrospectively get planning permission by claiming that they acted in ignorance and that the use is now so well established that it should be allowed to continue. This process asks the LPA to issue a Certificate of Lawfulness of Existing Use or Development stating that a use is *de facto* established and has thus gained planning permission by default. Potentially this process offers a back-door route to permission and it should be used only for genuine cases and where there has been no loss of amenity for the area and where the neighbours have not complained. The final right is a 'Determination' as to whether planning permission is needed or not. If a determination says that permission is not needed the applicant can go ahead with the proposal.

Summary

Development control is a very important activity, since it has shaped and continues to shape the everyday environment. Its key feature is that every application is treated on its merits but within a broad framework of guidance, notably the Development Plan and a raft of guidance from central government. Many decisions are straightforward, but some are complex and controversial. Each application does, though, go through much the same process of consultation, formulation of a recommendation and then a decision. There are, however, significant variations, depending on where the application is made, and in most areas there are strong presumptions against permission being given. The Development Plan should guide developers to where permission is most likely to be given. Whether this actually happens is the subject of Chapters 3 and 4.

Conclusion

Booth (2003) in his magisterial history of development control has praised the system for being the glory of the planning system because of its excellent processes. In particular, its responsiveness to change and its relative lack of legal complexity. However, the trouble with an emphasis on process is that it may obscure the ends which the process is designed to achieve. So, Booth concludes:

Modern development control is a palimpsest of attitudes to the urban environment, an accumulation of ideas and practices, which are often value laden, but which have often not been fully articulated. We have often given the appearance of not knowing what we want development control for.

(p. 24)

Accordingly, the book now proceeds in Chapter 2 to discuss the objectives of planning in the context of an evaluation system that places objectives at the initial stage of evaluation.

Further reading
in addition to reading highlighted in the text and illustrative material
The following texts will give further depth and detail to the readings already highlighted in the chapter.

Booth, P. (2003) *Planning by Consent: The Origins and Nature of British Development Control*, London, Routledge. A detailed history of the evolution of development control.

Cherry, G. (1974) *The Evolution of British Town Planning*, London, Intertext.

Cherry, G. (1996) *Town Planning in Britain since 1900*, Oxford, Blackwell. Both these books by Cherry provide detailed histories based on meticulous research.

Tewdwr-Jones, M. (2002) *The Planning Polity: Planning Government and the Policy Process*, London, Routledge. A detailed review of macro-level policy developments and local responses from the Thatcher period onwards.

Ward, S. (2004) *Planning and Urban Change*, second edition, London, Sage. This excellent book provides a detailed history with copious illustrations about the history of planning from 1890 to the 1990s. It provides the perfect complementary text to this one.

2 Issues in Evaluation

All human activities can be gauged a success or a failure. Sport provides a simple example, where the score provides a clear-cut definition. However, other activities are far more complex and measuring their success is beset by many problems. Thus most government policies, which axiomatically are complex, were not formally evaluated until recently. However, since the 1980s, state activities have been increasingly measured by performance targets under a political paradigm that demands 'value for money'. For example, schools are graded by exam performance, hospitals by the length of their waiting lists, and universities by the quality of their research and teaching. This has spawned a new pseudo-science of public policy evaluation (Howlett and Ramesh, 1995) with a range of evaluation techniques to satisfy the 'audit' culture which mistrusts the judgement of professionals.

Land use planning is one of the most complex state activities and accordingly its evaluation poses great difficulties. Indeed, this book is the first attempt to provide an in-depth evaluation of the planning system in Britain. In order to provide an evaluation, some general principles need to be discussed both in the wider context of evaluating public policy generally and in that of land use planning policy specifically. The central principle derived from work for the government by Morrison and Pearce (2000) is that there are links between objectives and evaluation with one flowing into another, as set out in Figure 2.1.

In more detail, all public policies must now have explicit *objectives*. These objectives will produce two types of *output*: first, a set of public policies, and second, a set of decisions based on these policies. These decisions will then produce an *outcome*, which will be a physical entity; for example, 150 houses will be built in a village. When the houses are built, there will be a series of *impacts*, for example, traffic flows will increase. When all the impacts are considered together, an evaluation can consider whether the policy objectives for land use planning in the village and the wider environment have been met using criteria set out in the original objectives. There are thus two types of evaluation: overall evaluations and those specific to outputs, outcomes and impacts.

This book uses all five links (objectives–outputs–outcomes–impacts–evaluation) as the key organising concept to structure our evaluation of land use planning. In particular, this approach is based on the classic method of arguing from the general to the particular and then back out to the general. It can be likened to an old-fashioned egg-timer in which all the sand starts at the top as the timer is started. The sand filters down through the funnel as specific issues are viewed through the waist of the funnel and then the issues are reaggregated as the sand falls to the bottom of the timer. This chapter concentrates on the factors at the top and the bottom of the egg-timer: objectives and evaluation, with some generic discussion of outputs as well. The key objectives and sections of the chapter are thus:

1   To understand why public policy is a compromise between the different value systems in a complex society like the United Kingdom and thus has limitations.

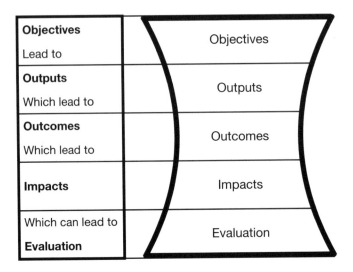

Objectives	Objectives
Lead to	
Outputs	Outputs
Which lead to	
Outcomes	Outcomes
Which lead to	
Impacts	Impacts
Which can lead to	Evaluation
Evaluation	

Figure 2.1 Flow chart from objectives to evaluation

2   To situate planning in the wider context of the society in which it operates in order to discuss whether it is a servant of the system or a catalyst of social change.
3   To discuss what other procedural options are available to the planning system.
4   To discuss the problems involved in evaluating planning.
5   To discuss the merits of various evaluation systems, notably a hybrid approach, and to outline the main form of evaluation used in the book.

Learning objectives

The main purpose of this chapter is to discuss the relationship between the political climate and the environment we live in, how the two are related by land use planning and what planning can and should do about managing environmental change. The main learning objectives of the chapter are thus to provide a contextual framework in which to situate subsequent chapters, notably the various types of evaluation in Chapter 5. The key point is that there is rarely one generally accepted view of reality, and indeed that reality is a concept that is constructed and contested by different groups dependent on their viewpoints.

PUBLIC POLICY WITHIN THE CONTEXT OF CONCEPTS OF POWER AND SOCIETY

Planning can operate only in the context of the society it finds itself in. It is thus essential to evaluate what type of society planning operates in and the way that power operates in that society. The next step is to evaluate whether planning is a prisoner of its host society or whether it can act like a cuckoo in the nest and change society for the better. Accordingly, this section examines the nature of power and society and planning's various roles and objectives in managing environmental change for society.

Power

Political power is exercised by the state via politicians and public-sector workers. Politicians are elected by the people and carry out a legislative programme broadly based on ideology tempered by the pressure of day-to-day events and opportunities. A few public-sector workers at the strategic level help to develop policy while at the executive level power is delivered by thousands upon thousands of workers. This is, however, a benevolent view of power in which individuals willingly surrender freedom to the state and its agency the government, which governs by consent. It also represents one end of a spectrum, which terminates at dictatorship. In between, the most realistic models explain power being exercised in either a consensus or a conflict-ridden society, as exemplified by the work of Habermas and Foucault.

Habermas (1984) and his disciples in the school of critical theory hold a liberal view of power in which knowledge and language are used together by different groups to achieve consensus via the concept of 'communicative rationality'. The basis of this approach is a democratic context in which anyone may question the claims of another, and so in order to provide social justice, governance structures must seek out the views of those currently excluded. Habermas suggests that there are five tenets to such an open and democratic debate. First, the debate must be genuinely open to all. Second, all actors should be autonomous. Third, all parties should respect and indeed be able to empathise with each other. Fourth, all parties should be honest and open, and fifth, they should thus abstain from hidden agendas.

In contrast, according to the influential views of Michel Foucault, power does not exist as an entity, but is exercised through a network of social relations, which are inherently unstable and changeable. This is because there is resistance to the networks of power and because the knowledge which gives power its legitimacy is a constructed yet contested discourse. Power is therefore transitory and depends on a network of social relations that recognises one discourse to be the 'valid' truth at any one time. Thus, because power is not an object, we can only study the specific social practices through which power relations are put into practice, because power exists only when it is put into action (Foucault, 1982). A Foucauldian analysis of public policy would divide power into five themes. First, dividing areas up into different policies. Second, examining the different goals of different actors. Third, examining the different means used to put policies into action. Fourth, the institutions that are created. Fifth, how effective the power has been in achieving results and the costs involved. These five themes according to Murdoch et al. (1999) are not very different from the views of Habermas, in terms of power structures, except for the crucial difference that power and knowledge are seen to be constructed and contested in ways that are inimical to achieving consensus.

In addition to these two key concepts of how power operates there are several other differences in how the exercise of power can be explained or executed: different concepts of the state, spatial considerations, and the power of personality and chance events. These are now considered in turn.

Different concepts of the state. The traditional view of the state in Britain has been based on the 'pluralist' model in which each individual or group has equal access and influence to those in power. More realistically, an 'elitist' model argues that some individuals or groups have favoured access and unequal influence, for example developers. A variant of the 'elitist' view, the 'corporatist' model, contends that organised groups have the greatest influence, notably those groups who are thought to form the 'establishment', for example Oxbridge-educated civil servants or professionals. A more radical view is based on Marxist rather than orthodox economics and argues that only the interests of capital have any influence, and that even the state is subservient to the imperative to accumulate capital.

Spatial considerations. We live in an increasingly global world, and power is now exercised extra-nationally at the global, continental and international levels. Nation states, including Britain, have thus had to surrender certain powers to these extra-national organisations. Thus planning has to be aware of these wider forces, most notably global capital flows, although in terms of legislation, planning is still very much in the hands of the UK government. Within the United Kingdom, there have been increasing calls for both federal and regional systems of government, for more decisions to be made at the local level and for planning to take an integrating role. This used to be controversial in a country like the United Kingdom, which has traditionally employed a top-down system of power in which legislative decisions are taken at the centre and then executed with little or no discretion locally. The principle of subsidiarity is crucial here. Its core idea is that the overall framework in which decisions are made should be set centrally, but that the detailed execution should be decided locally. In essence this is encapsulated in the slogan 'Think globally, act locally'.

The power of personality and chance events. Even with apparently robust systems powerful personalities can still change the course of events against not only the opposition of opponents but even supposed allies. Thus, for example, in the 1980s Mrs Thatcher (the iron lady) was able to move not only the goalposts of politics but even the playing field, even though she never commanded more than just over 40% of the votes cast in her term as Prime Minister. Similarly, chance events can have a great influence. For example, in the early 1970s a crisis in energy supplies brought about by war in the Middle East and striking miners in the United Kingdom effectively brought an end to a style of planning based on inner-city bypasses and road building.

Public policy in Britain has grown incrementally as Parliament has gained more and more freedom from the Crown. The British democratic system has six key features. First, the government nearly always represents less than 50% of the votes cast and sometimes only 30% of the electorate. Second, the system is adversarial. Third, legislative change is slow, for example planning gets a major new Act every 15–20 years and the 2004 Planning Act took over 18 months to pass. Fourth, Acts of Parliament are usually preceded by several formative policy statements, known as Green or White Papers, green for policy options, white for draft policies. Acts of Parliament are only bare skeletons couched in vague terms. Detailed policies are evolved via speeches, circulars from Ministers and detailed legislation, which provides the organs, muscles and flesh in the form of Statutory Instruments. (For detail see GPDO in Chapter 1.) Legal interpretation provides a further stage of fleshing out the legislative bones. Fifth, the system is based on privilege, with personal contacts often being crucial. Sixth, the system is sectoral and based on developing policy in discrete areas rather than seeking holistic solutions. A lot of agencies operate the system via constructive or destructive tension. Planning is of course an attempt to provide a holistic view but planning policies are also often divided up sectorally, for example policies on housing, transport and employment.

In summary, the process is thus incremental, accretionary and somewhat haphazard in its evolution, depending on the political party in power and the length of their Parliament. This means that we have to adopt a pragmatic rather than a conspiratorial view of planning history.

Land use planning has been broadly accepted by all political parties since its inception in 1947, but as Box 2.1 shows there have been six main political epochs in post-war Britain. The overall trend apart from the 1960s has been the 'privatisation' of planning as a service to enable economic growth to take place without too much damage to the environment.

Box 2.1   Political history of post-war Britain

1945

Election of Labour government with a radical agenda to create a 'welfare state' based on state provision of housing, education and health services and control over most of the means of production of core industries.

1951–64

Long period of Conservative power under which the private economy flourished and people 'never had it so good'. But in spite of this prosperity Labour was returned to power after a number of scandals and the belief that the Tories were out of touch as a party run by patrician rural landowners in a country becoming increasingly urban.

1964–70

Labour under Harold Wilson tried to create a planned society based on merit, education and science.

1970–79

The Conservatives under Edward Heath had disastrous confrontations with the coal miners and were replaced by Labour in 1974, who lacked the energy and vision of the 1960s. The power of the unions was a key factor in resisting change and Britain fell behind its industrial competitors.

1979–97

Margaret Thatcher elected on a radical anti-state platform and many features of the post-war consensus were swept away, notably the power of the unions (the coal miners' union was crushed) and many state industries were privatised. Core principles of the private sector, e.g. performance targets, introduced into a much reduced state sector, albeit containing two major employers: education and the NHS.

1997 to date

Tony Blair elected on a neo-Thatcherite platform which had already been modified by John Major (1990–97). A continued belief in the efficiency of the private sector as a wealth creator but a growing belief that the private sector needed to be controlled and taxed to provide better health and education services, which had been relatively run down under the tax cuts of the Thatcher/Major years.

Society

Political power is said to be exercised for the benefit of society, but society is a word that has many meanings. Three dictionary definitions stand out. First, the customs and organisation of a civilised nation. Second, a social community in which no society can retain members who flout its principles. Third, an association of persons united by a common aim or interest or principle. Society thus implies a state of semi-equilibrium where people have forgone some individual freedoms and desires in favour of a greater common good upon which all can agree, as proposed by the influential philosopher John Rawls (1993). Crucially, in doing so they have surrendered many powers over their own destiny to organisations that can impose their own norms and regulations. Thus, although a society may be in reality a cauldron of conflict, the fear of punishment by the state or retribution from others prevents most of us from departing from legal and cultural norms, and long-term self-interest encourages us not to act in short-term antisocial ways. More positively, the influential school of 'communitarianism' has developed from both empirical evidence (Putnam *et al.*, 1993) and from philosophical first principles (Etzioni, 1993). Its key point is that the performance of institutions is shaped by the social context in which they operate, and that beneficial contexts are provided by a high degree of mutual trust, social co-operation and a well developed sense of civic duty.

The concept of society is thus crucial to understanding planning, since planning has to find common value systems around which to base its policies, and the degree of compliance or deviation from these policies that it can expect. In order to gain acceptance of these policies planners will also need to gauge the correct mix between self-interest incentives and retributive sanctions that it will employ.

Value systems within society

The environment and planning are both social constructs based on individual value systems. The key point to be made here is that traditional value systems based on a common religion and moral philosophy and the UK legal system based on Judaeo-Christian ethics have both become increasingly challenged as religious belief has fallen away. The main movement has thus been one away from absolute beliefs, as typified by the Ten Commandments, to one of relative beliefs which has been labelled 'permissiveness' by its critics and 'political liberalism' by its advocates (Rawls, 1993). Planners are thus faced with an increasing variety of opinions about what the environment should be, and how it should be managed. Traditionally they have taken the side of ruling elites and focused their policies through the lenses of developers and landowners, especially in a country like Britain where land ownership still confers great social status and owning a house is seen as very desirable.

But in recent years environmental groups ranging from the traditional to the radical have entered the scene, and while most are happy to employ peaceful forms of action, a radical fringe is willing to embark on 'direct' action. Planners can either celebrate this diversity of viewpoint or, more traditionally, they can try to seek consensus. To do this their favoured technique is 'public participation' by which they seek to ascertain the 'public interest'. Even though as Thomas (1997) argues the 'public interest' is elusive to define it is nonetheless deemed to be the apparent cornerstone of day-to-day decision making.

Taylor (1998) has thus identified two ways by which planners can define the 'public interest'. First, the Benthamite concept of the greatest good of the greatest number, based on the work of the nineteenth-century philosopher Jeremy Bentham and then developed by John Stuart Mill into 'utilitarianism'. Second, the Hegelian view of a nation or society subscribing to some overall ideal or purpose, for example the setting up of the welfare state in the late 1940s as a reaction to the failures of free enterprise in the 1930s and the privations of the Second World War.

Summary

Land use planning operates within the context of a political system that is dominated by a few motivated individuals who govern us largely because of our apathy. Policy changes are incremental and unlikely to be controversial. The key ideology within politics since the 1980s is that the state should not interfere with the private sector unless necessary. Therefore, public policies need to be justified in some way if they are to be successfully implemented, and so attention is now turned to reasons why the state should interfere with the freedom of the private sector, with particular reference to land use planning.

WHY SOCIETY MAY WANT PLANNING TO INTERVENE, WITH WHAT OBJECTIVES AND BY HOW MUCH?

The decision by the state to intervene depends as much on the ideology involved as it does on the issue. Ideologically left-of-centre or centre parties like the Labour Party or the Liberal Democratic Party are instinctive interveners, while right-of-centre parties like the Conservative Party are instinctive non-interveners. Within this framework, however, all politicians have four basic options, according to Delafons (1995) in his reflections on his decades as a civil servant dealing with planning. First, do nothing or pretend to do something by setting up a committee and hope that the issue will go away or divert the blame elsewhere. Second, do the bare minimum to deflect public criticism. Third, appropriate the policies of the opposition and steal their thunder; fourth, make a virtue of a problem and actively espouse intervention as a good thing, thus turning defence into attack. In actual practice the most common option is a fifth one, modify existing policies by gradual incremental change, usually only in response to a crisis.

Thus policy making can be seen as a process of crisis driven by accretionary incrementalism, which produces a chaotic set of increasingly complex and contradictory policies. Delafons (1995) also points out that bureaucrats rarely propose policy changes, since they are far too busy implementing current policies, or clearing up old ones. Furthermore new policies mean more work. Therefore the decision to intervene must be backed up with powerful arguments as to why there should be intervention, what the objectives should be, and then how much intervention there should be to achieve these objectives. These issues are now considered in turn.

Why do we plan? To reconcile different value systems in society

Land use planning is needed for two compelling reasons. First, to resolve conflicts over land use, and second to allow the public to have a say in those conflicts. There is much agreement that the main aim of planning should be to balance the competing demands of society, the economy and the environment, as shown in Figure 2.2. There is, however, a vigorous debate about what that balance might be, depending on different viewpoints about what is important. Planning is fundamentally about land use change, and although many people would like it to be more than that, the legislation deliberately restricts it to land use issues. Its main purpose is to resolve conflict and provide an arbitration service in regulating change. In particular it

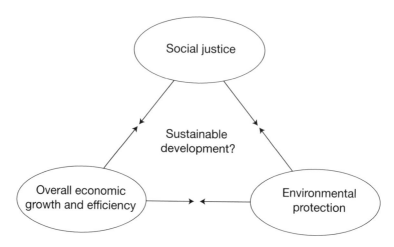

Figure 2.2 The triangle of conflicting goals for planning. *Source:* adapted from Campbell (1996)

seeks to maintain equilibrium between social and economic needs and environmental protection. However, it is expected that new jobs and houses will be allowed by planning unless the environmental objections are significant.

Take for example Heathrow airport. Planes descend on precisely ordained flight paths every 40 seconds, thousands of meals are served every day, thousands of people arrive and depart every hour and yet there are surprisingly few long delays. It is a miracle of organisation and intensively planned but the potential for gridlock is never far away. Should it grow any more or not? This was the subject of a 500+-day public planning inquiry in the 1990s, which considered a mass of conflicting arguments. For example, noise and pollution for residents, economic growth in the South East compared with other regions and so on. Finally, it was decided to expand the airport by an extra terminal but to keep the same number of flights, by using bigger planes and cutting out smaller planes. This was a classic balancing of the demand of society and the economy for travel and the needs of the environment.

All forms of public planning as opposed to planning by individuals or private-sector organisations imply the intervention of a third party into current activities. It is thus axiomatic that planners have to tread a delicate path between alienating the overarching state of which it is a part, by taking over too many of its powers, or alienating the public who have sought their intervention. In addition, planning powers cannot be so onerous that they prevent reasonable activities from taking place, or be so lenient that they may as well not exist. Whatever the form of public planning, however, the planners will find it hard to win, since they are trying to impose the concepts of a common good on a diverse and plural society. In particular, according to McAuslan (1980), they are trying to reconcile four competing ideologies: private property, the public interest, democracy and justice. McAuslan suggests that one of these will be dominant at any one time. If that is so then attempts to achieve reconciliation will be controversial, often unpopular and will be resisted, so to survive planners must continually justify intervening where others may fear to tread.

Conflicts over land use are exacerbated by the issue of land ownership, which in Britain has traditionally bestowed rights and privileges on the landowner. Quite reasonably landowners wish to use land as they see fit, but this may harm the amenities of other landowners or land users. In addition, third-party interests may object to the use of the land on grounds of pollution, waste and ugliness. There is thus a conflict between freedom and control. By and large British society lets the free market decide on many issues, but the market fails to take into account what economists call externalities, for example the cost of cleaning up pollution. In addition the market does not place a price on so-called 'free goods' like fresh air and fine landscapes. Collectively these are known as 'market failures' and they provide the main justification for planning to intervene in land use management.

But society is in reality a cauldron of conflict and tensions. Thomas (1997) argues persuasively that capitalist society is postulated on competition and conflict. Accordingly, Thomas (1997) has outlined four reasons why people need planning:

1 To resolve conflict.
2 To provide democratic and accountable decisions.
3 To reconcile private and public interest.
4 To provide equity and fairness.

Strangely, as Alder (1989) has argued, planning has no purposes of its own, only as a microcosm of society. As such it has aggregated a number of purposes and value systems, notably utilitarianism (the greatest good of the greatest number), paternalism (Daddy/Mummy knows best), utopianism (an ideal world) and the protection of property rights (Keep out! Beware of the dog!). The lack of any legislative purpose has tellingly been commented on by the report on planning by the Royal Commission on Environmental Pollution (2002) but the 2004 Planning Act did for the first time spell out that the purpose of planning was to create a sustainable society.

In the area of land use planning a spectrum of reasons can be produced, with each sector of the spectrum leading to a different set of planning powers.

Objectives for planning from different perspectives

From the literature

There are probably five times more words written on what planning should do than on what it has done. Indeed, the main function of this book is to remedy this defect, although I myself am guilty of adding to what I call 'must do' publications with my book *Countryside Planning Policies for the 1990s* (Gilg, 1991). This trawled over 1,000 proposals for policy changes or 'we must do this' type of rhetoric and demonstrated the enormous number and diversity of possible planning objectives. A later example is provided by Thomas (1997), who has aggregated the purposes of planning into the following 12 objectives:

1 To balance protection of natural and built environment with pressures of economic and social change.
2 To set land-use framework for economic development.
3 To make things happen at the right time and the right place.
4 To meet need for housing.
5 To facilitate accountability.

6   To protect landscapes.
7   To control the external appearance of developments.
8   To secure efficient use of land in the public interest.
9   To resolve conflicts.
10 To act as a forum for public information.
11 To be mindful of global, international, regional and local issues.
12 To ensure sustainable development,

From the government

In more detail the government sets out specific objectives for planning from time to time as shown in Box 2.2. These are broadly the same from decade to decade but priorities change and new objectives, like sustainability, come to the fore. An instructive exercise is to allocate these functions and objectives to the society–economy–environmental objectives shown in Figure 2.2. This is a difficult task because many of the functions and objectives overlap. For example, is the housing function an economic or a social one? Of course it is both, since house building is a major wealth creator and people need houses to live in. But the type of housing will be ordained by which objective is more important. It is this ordering of priorities that is the main task of planning. In other words, in any one case which purpose is most important? Equally important, though, is how much intervention is needed if these objectives may be achieved, so this issue is discussed in the next sub-section.

Box 2.2   Government objectives for planning in 2002

In 2002 the relevant planning Web site was the Department for Transport, Local Government and the Regions and its home page gave the DTLR five functions: Transport; Planning, cities and communities; Housing; Local government and the regions; Health and safety. Planning was defined as:

> Promoting a sustainable pattern of land use, through an efficient planning system. Improving the quality of life for everyone in our towns and cities and renewing the most deprived communities.

Following the link to Planning produced different definitions:

> Our aim is to improve the quality of life for everyone in our towns and cities and to renew our most deprived communities so that nobody is seriously disadvantaged by the place they live.

> An efficient planning system promotes a sustainable pattern of land use. Planning promotes local economies, regenerating rural and urban areas and protecting our valuable heritage.

And according to the link to the December 2001 Green Paper on the future of planning (see Chapter 5):

> an efficient planning system is the key to delivering the government's wider sustainable development objectives.

Thus the key words/phrases seem to be: sustainable land use, efficient delivery systems, quality of life, deprivation and regeneration, local economies, and heritage.

Of course there are other government bodies that influence these aims, and so visit current Web sites for these organisations, notably DEFRA and the Environment Agency.

At the time of writing (July 2004) planning had come under the control of the Office of the Deputy Prime Minister with the Web address www.odpm.gov.uk.

How much should we intervene? A spectrum of views

As we have already seen, plan making is a fundamental trait in the human pysche and in mixed economies like the United Kingdom society accepts the need to impose controls on the way in which resources are used. In a small and crowded island like Britain, land is one of the scarcest and most fought and thought over resources. Therefore there is a powerful case for a system that attempts to manage land so that the right use takes place at the right place and at the right time. Just what is right is of course a matter of often bitter dispute and thus there is a clear need for a system of arbitration between conflicting parties. The conflicting parties have different resources at their disposal and so the system will need to appear to treat people and organisations equally and to find out what they want from the environment. The justifications for intervention in the market are, though, a matter of debate but three broad viewpoints can be discerned along a spectrum from little planning control to virtually complete control based on O'Riordan's (1985) spectrum of environmental ideologies ranging from ecocentrism to technocentrism. The three main viewpoints in this spectrum are now discussed in turn.

Virtually complete control, since planning is essential

These are the arguments spelt out by communists and deep-green ecologists. Communists argue that planning should promote social justice, preferably by transferring productive activity to communal ownership. This view is not widely held at present, but it has been influential in Marxist/totalitarian regimes. It was the mainstay of Britain's war effort in 1939–45, and was thus influential in setting up the 1947 town and country planning system.

In contrast deep-green ecologists predict that planning of a draconian sort is needed to prevent eco-catastrophe. There are two main problems with this view. First, there is a long lead time between misuse of the environment and its impact. For example, the greenhouse effect on climate change cannot yet be totally attributed to increasing burning of fossil fuels. Second, pollution and resource depletion are no respecters of frontiers; for example, the acid rain that plagues Scandinavia is caused by pollution from Britain. Thus there is little point in one country going alone. Third, deep-green politics in a free mixed economy are basically non-electable, since car owners are not easily going to vote for restrictions on use any more than turkeys would vote for Christmas. Thus, pragmatically, deep-green arguments cannot make headway, except in creating a 'conservation culture' based on keeping all environmental options open and the precautionary principle: If in doubt, don't.

Planning needed to guide and shape a mixed economy: reactive or proactive?

This is the core argument and is broadly accepted but it contains two differing roles. The first is reactive and arbitrates between developers, both public and private, and the rest of society. The second is more proactive and tries to resolve conflicts by bargaining and by imposing conditions or even more proactively by preventing conflicts by forecasting land use change and allocating land by zoning. More ambitiously planners may try to create places by better design, attempt to manage macroeconomic trends like the move from an industrial society to a knowledge-based and distribution society based on consumption, not production, and revive city centres by attempting to provide a 'café culture'.

The reactive role: planning arbitrates between developers, both private and public, and the rest of society in its various groupings

The arguments for a reactive role are focused on making the market more efficient or by correcting market failures. There are two different ways, however, by which market failure can be addressed, according to Webster (1998). The traditional welfare economics approach lets the government decide on the way in which social objectives should adjust to market failures, while in contrast the more modern form of public participation planning is expressed by voters or pressure groups demanding government action on market failures. These latter arguments centre on the market's failure to take environmental goods and losses (pollution) into account, namely the 'externalities' of any activity. In the British context arguments include the fact that Britain is a small, crowded island with a high and growing degree of conflict between land uses, most notably between urban and rural interests and between food producers and consumers. In the wider ecological context, arguments range across varying shades of 'greenness', as shown by O'Riordan (1985). Other typical arguments include: seeking consensus; resolving existing conflicts; preventing possible conflicts; balancing economic growth with the preservation of the environment; balancing efficiency with equity; managing change; co-ordinating development; and seeking the greatest good of the greatest number (utilitarianism). Ecologically, key arguments include the 'the polluter pays' principle and the concept of the best practicable environmental option.

All these arguments can be conflated into the triangle of conflicting goals shown in Figure 2.2 in which ideally the economy, the environment and society can be reconciled into green, profitable and fair 'sustainable development'. This is of course difficult to do, not only because each corner of the triangle wishes to maximise its own interests, but also because individuals act differently in different scenarios. For example, in the classic pedestrian or car driver scenario individuals act differently depending on whether they are in the car or not. Thus the planners are in a no-win situation:

1   If development is refused planners are seen as the enemy of economic growth and jobs.
2   If development is always permitted planners lose the rationale of their existence.

One way out of this dilemma is to pursue the second, proactive role.

Proactive role: to resolve conflicts by bargaining and by imposing conditions, or if possible prevent conflicts by forecasting demand and allocating land rationally by zoning

The arguments for a proactive role include putting the community in control; making things happen that wouldn't otherwise and making things better (utopianism), which was one of the arguments of the founding fathers of town and country planning. A simple example is provided by pedestrianisation, which has massively improved our environment by taking noisy polluting cars out of our High Streets and allowing us to walk freely among assorted vegetation. Planners can also use trade-offs or the concept of 'planning gain'. A trade-off occurs when one party to a dispute forgoes certain advantages in order to achieve the desired outcome. 'Planning gain' is the converse, when planners seek some public good from a developer as a necessary requirement for granting planning permission. Both processes open up the possibility of graft and corruption, so it is important that procedures are in place to make these processes either transparent or open to scrutiny. The use of free-market concepts leads us to the right wing of the spectrum, the Conservative Party view that planning is a necessary evil.

Planning is a necessary evil

The arguments for this role focus on the need of business to have freedom but not a freedom that is self-defeating because as Thomas (1997) has argued business needs some sort of controlling authority in order:

1   To co-ordinate investment in land and buildings.
2   To ensure development works in a functional sense.
3   To ensure development is visually pleasing.
4   To fulfil the vision of development plans.

Even Mrs Thatcher when she had been in power for one year, and had viewed all the possible targets in the public sector for either privatisation or elimination, had to conclude in DOE Circular 22/80 that 'The planning system balances the protection of the natural and built environment with the pressures of economic and social change. The need for the planning system is unquestioned and its workings have brought great and lasting benefits.' This is, however, a negative view of planning as something that prevents bad things rather than promoting good things. Key examples are severe instances of market failure: preventing profiteering from land; preventing chaos; preserving property rights/values; protecting the poor and the weak from exploitation; and preventing poor landscapes being created.

In spite of planning attempting to remedy market failure, many critics point to planning failure and claim that the end result is no better. For instance, high levels of traffic congestion, the inflationary effect on land values, delays causing a loss of economic activity and jobs, poor levels of design, the destruction of old buildings and communities, run-down housing estates, characterless suburbs, bleak city centres and sprawling out-of-town commercial activity. Another critique is that planning infringes the liberty of the individual and that some world-famous cities like Edinburgh were built using different methods, for example design competitions and covenants. On the wider scale, attempts to redistribute wealth and economic activity by regional planning have not been very successful. Public transport has failed to match the attractions of private transport and planning has failed to pay enough attention to energy saving either in the design of buildings or in their location (Breheny, 1995).

At the end of the spectrum is the view that planning decisions should be left to market forces because the concept of public planning is inherently flawed. For example, Pennington (2002) argues that planning is too bureaucratic and does not allow experimentation. In addition, planners cannot make decisions in the public interest because they are subject to pressure from interest groups, which are lobbying against development, thus creating land shortages. Instead, Pennington argues, a private system of land use control based on covenants, deed restrictions and the establishment of 'proprietary communities' would provide a way 'to hit people with the consequences of their decisions'.

Finally, critics of planning as an essentially future-based activity have argued that it is not possible to predict the future with any great accuracy. Planners are too often forced to retreat into so-called trend or 'predict and provide' planning where they merely track current trends into the future and allocate land to the activities predicted. Critics point out that this produces self-fulfilling prophecies that axiomatically render planning redundant.

Summary

The central theme of all the reasons discussed above is that planning will make things better and that a consensus society is better than a confrontational society. If we consider two aspects of our own lives from the point of view of common sense we should be able to confirm a gut feeling that planning is a 'good thing'. First, let's consider how we act individually, and second, how we interact with others. As individuals we all plan our lives to some extent – when to eat, when to sleep, what career we want and so on. But as individuals we interact with other individuals, and in order to do so we need a set of informal and formal codes of behaviour, and notably some agreed set of goals. Thus to complete a degree course, there will be set times for lectures and for exams. The alternative would be chaos, and so the prime reason for planning is to achieve some sort of order. This section has discussed why and how planning might meet the goals of society.

REVIEW OF ALTERNATIVE DELIVERY SYSTEMS FOR PLANNING

Working independently of each other, Gilg (1991) and Selman (1988) have produced a spectrum of delivery systems working upwards from voluntary methods to increasing levels of control via incentives:

1 Voluntary methods based on exhortation, advice, demonstration, but often backed up with the threat or promise of one of the four methods outlined below.
2 Financial incentives to encourage production and/or desirable uses.
3 Monetary disincentives to discourage production and/or undesirable uses.
4 Regulatory controls, mainly negative, for example planning permission.
5 Public ownership or management of land via long-term leases.

These methods range across the political rainbow but they are not mutually exclusive and they can be combined. In addition to these powers there are three other options. First, set up some form of administrative body. Second, set up special areas, in which the powers are modified, for example National Parks. Third, use market methods to sell the right to develop land, as for example is widely practised in the United States.

In Britain the traditional process has been to start with the weakest power available and then to proceed up the regulatory spectrum as and when needed. The only exception was the Labour government of 1945–51, which introduced the entire spectrum in a series of landmark Acts, notably the Town and Country Planning Act 1947, the Agriculture Act 1947, the New Towns Act 1946 and the National Parks Act 1949. In land use planning the dominant form of delivery has been regulatory controls backed up by formal guidance in the form of a land use plan which indicates where certain types of development will probably be given permission. British land use planning is thus based on two key processes. First, the production of land use plans and policies, and second, decisions on applications to change land use based largely on these land use plans and policies.

In more detail, the favoured system of policy delivery has been the so-called *rational comprehensive linear systems* approach in which planners portray themselves as scientifically perfect, value-free, neutral technicians who present decision makers with a series of calculated

1   Planners should **think** about the problems and opportunities represented by the current situation by carrying out **surveys** and subjecting the findings to careful **contemplation** based on a **classificatory** analysis which **groups** and **ranks** the issues. The public should be aware, however, that planners are only human and may not always examine every aspect of the issues, through either laziness or prejudice, and that even when they use their abilities and techniques to their best advantage these may still not be good enough for the complexity of the issues involved.

2   These issues should then be extrapolated into the future derived not only from how planners **expect** the future to develop based on a number of **assumptions** but on how other groups may forecast the future depending on their **aspirations or concerns**. These different **aspirations or concerns** will cause **conflicts** over who gets what and **planners** may need to **resolve** these by a series of **resource allocation decisions**. Such **decisions** may also be avoided by **anticipating** future developments and if possible **preventing problems** by employing **the precautionary principle**.

3   One of the goals of planning is thus a reactive approach to problem-solving which is rooted in the Enlightenment idea of progress in general, and in progress via the scientific method in particular. Planning is not, however, just about problem-solving and reactive policies can be matched by pro-active proposals which make good things happen that wouldn't otherwise. Both types of policy are more likely to be **acceptable** if they are based on a **vision** of how things may be **better or worse** if the policies are not employed in the future by setting goals used on certain **beliefs** and seeking a **balance** between different groups on the 'greatest good of the greatest number' principle or put simply **common sense**. **Beliefs** can best be **balanced** if they are **ranked** into an order of priority. Planner's **explicit motives** can thus also be **ranked**.

4   Proposals for possible methods for achieving these **explicit motives** which include **resolving and preventing conflicts** may involve **exhortation via mediation, arbitration and reconciliation**, but some degree of **intervention** will probably be necessary and **plans** may often be best implemented by using a **regulatory approach**. When assessing possible **regulatory** methods for achieving stated **goals** planners can use their 'superior' **judgemental skills**, but they should also consult widely. The principle of consultation can also extend to reaching agreements which will be most effective if they are legally binding agreements. All planning processes should also identify resources that need **protection** and methods for **preventing** damage to these resources may include **site protection** and employing the core principle of **conservation/preservation**, good **husbandry**. The environment may best be protected, however, by reducing consumption of resources by exercising **thrift**, or 'making do and mend'. This, however, is a concept which planners may find hard to get across in a society dominated by built in obsolescence. In assessing which **methods** to employ planners should be **cautious** and set **realistic targets** based on what is **achievable** given their role in society, which may in essence be one of **co-ordination** only.

5   Whichever methods are chosen an organisation will need to be set up which can institute procedures for implementing the chosen methods which will need to be set out in a series of **plans, schemes or designs**, following a process of plan-building based on consultation, in which a spectrum of planning powers from exhortation to coercion have been outlined. This spectrum should contain three sub-elements and three main mechanisms as shown below:

Indirect influence	**(leadership, adjustment, guidance, codes)**
	Exhortation
	Financial incentives and Disincentives
Direct influence	**(government, rules, directions, precepts, decrees)**
	Regulations backed up by sanctions
Direct control	**(conducting, orders, commands)**
	Direct control or ownership of recourses

These mechanisms need to contain measures to ensure that they are not **exploited** or ignored and a set of sanctions and enforcement procedures need to be set up to police them.

6   The performance of these **plans** should be **evaluated** to assess who or what is **winning and losing**, and **planners** should set up a series of **performance indicators** by which their own work and their **plans** can be appraised. This **appraisal** should accept that there will be an implementation gap and should thus set acceptable levels of effectiveness in **achieving** the **goals** of the plan. In both cases the appraisal should attempt to find out why any **deviation** between the **plan** and reality has occurred, and propose **amendments** to the plan. This process will probably need to go on for ever and **planners** will have to be **externally vigilant** about new threats to the environment.

Figure 2.3   An ideal planning system? *Source*: Gilg (1996)

options, as shown in Figure 2.3. The underlying assumption is the 'utilitarian' one that the best option will be the one that will entail the greatest good of most people and the least harm to the smallest number of people. The main vehicle of delivery should be an end-state plan representing the utopia to which the plan aspired. Since the 1970s, however, a number of other methods have been advocated. In particular, public participation became more important, and the degree to which planners could be seen as value-free technicians began to be widely queried. Planners have reacted by putting forward more alternatives and making much clearer both the technical and the judgemental basis on which their proposals were founded in an espousal of postmodern uncertainty.

An ideal system can nonetheless be predicated and indeed it would be in the nature of planning to strive for such a system. Such a system would set out the desired goals, the negative factors preventing those goals being achieved and the positive factors aiding those goals being achieved, the potential for eliminating negative factors and for enhancing positive goals. Finally the system would set the percentage of failure that would be acceptable before a policy review would be needed. The system should also set out the policies in a statement and/or maps and diagrams demonstrating how the policies will work, who is affected and how, and what they will need to do, if anything, as shown in Figure 2.3.

An ideal system would also note the problems likely to be encountered in practice. First, the problem of overlapping time horizons between past and present policies, differential priorities between different goals, and overlapping spatial scales. The system would also need to take into account the workability, efficiency and affordability of the policies, the degree to which they are undermined by partial knowledge of the present and uncertain predictions about the future. Finally, the likelihood that the policies will be blown off course by unforeseen or chance events in the future will need to be considered.

In an overview of alternative planning methods Alexander (1998) examined four overlapping theories. First, planning as deliberative rational planning carried out by planners. Second, planning as interactive rational planning with actor networks. Third, planning as co-ordinative planning but with individuals rather than organisations, and fourth, planning as frame setting, working with communities to build consensus. This fourth model has been termed 'communicative planning' and then developed into the concept of 'collaborative planning by Healey (1997) examined more fully in Chapter 5.

A more complex typology of planning methods has been developed by McDonald (1989), based on the classic work of Friedmann (1973) which demarcates four styles of allocative planning. First, *command* planning, which sets compulsory targets; second, *policies* planning, which structures decision environments; third, corporate planning, which uses processes of bargaining and negotiation; fourth, *participant* planning, which relies on voluntary compliance, with preferences reached in group deliberations. In a critique of how these four styles may be delivered in practice McDonald rejects the rational planning approach as too simplistic for a complex world of conflict, except where narrow sectoral plans are involved. McDonald also rejects the conflict resolution role because there is no consensus among policy makers. Thus McDonald advocates the role of planning as bargaining.

Sager (2001a, b, c) has attempted to link the management styles of planning agencies and the ethos of their staff with a series of planning styles in an attempt to show that the rational planning/procedural planning approach can be justified. Sager in an argument against the critics of rational planning/procedural planning concludes that social choice methods can indicate to planners what their main policy directions must be, but that within these broad choices they can manipulate policies according to the type of culture operating within their agency. Thus the agency is given a mandate by general public consent, the management of the agency then introduce a management style based on trade-offs between different goals biased towards their beliefs.

At the day-to-day level individual planners also use their preferences, and over a period of time there is congruence between the structure of the agency and the decisions of its officers which produces a plan or set of policies which reflect the style of the organisation. Depending on the style of the planning agency, Sager argues, they have a choice of four planning styles: *synoptic* (technocratic) planning; *incremental* planning, which accepts uncertainty; *communicative* planning, which attempts to involve the public; and *advocacy* planning, in which planners advocate policies both within and outside their agencies. The degree to which planning policies follow the rational planning model thus depends on the management style of the planning agency and its subsequent choice of planning style. In conclusion Sager argues that though planners can manipulate planning policy the inherent uncertainty in attempting to plan the future will always cause great difficulties.

In essence two broad delivery systems can now be discerned in practice. First, variations on the rational cyclical process system model in which the planner is dominant. Second, an interactive model in which planners act with key groups. In either model the decision-making process can take two forms. First, an incremental crisis response model limited by imperfect resources of time, knowledge and political realities; second, a muddling-through managing-change model. The net result is that planning is now seen as a decision-centred exercise among wider groups in society, with ideally the planner being seen as the centre of a large grouping of corporate bodies. In this role, planners make checklists of how developments or plans may affect their area and then make a decision which causes the least harm in light of the views of key groups. In particular, planning is now seen as a bargaining process between key 'stakeholders' in which planners may often take a leading role.

Bargaining is in fact the main delivery system that has emerged since the 1970s, based on the plan as a framework only, which is then employed only as a starting point in negotiations between interested parties. The main activities are thus seen as bargaining, with planners acting as an arbitration service between parties and thus attempting to resolve or prevent conflict. The essential process is one of decisions made on the merits of each case set within only loose overall guidance.

Summary

All public policies can be placed on a spectrum from exhortation to complete control. British planning mainly uses the second item on the spectrum, regulation, but initially it used the fifth item, public ownership and development of land. Regulation can be delivered in a number of ways but the rational planning model has dominated in which plans for the future are based on apparently scientific evidence and forecasting. However, day-to-day regulation is more pragmatic, is based on bargaining between key groups and thus involves a good deal of discretion and flexibility. This discretionary element in decision making is one of the main problems in evaluating policy, particularly the supposed link between policy and implementation. Nonetheless, the 'audit' culture demands that planning, in line with all other public services, evaluates itself and so the next section discusses attempts to provide an evaluation framework.

Summary of nature and purposes of planning

Carmona *et al.* (2003) have provided a pithy review of the nature and purposes of planning, in which they note that planning is multi-dimensional, multi-objective and concerned with the complex management of change in the built and natural environment. They then adapt work done by the Quality Assurance Agency for Higher Education (2002) to benchmark the nature and principles of planning for students of planning under six headings. Carmona *et al.* then adapt work by Rydin (1998) to provide a list of three rationales or purposes for planning under the 'six headings' shown in Box 2.3.

Box 2.3 Benchmarking planning

1. Planning is concerned with relationships between society and space.

Planning is about determining the quality of the relationships between people and space. Planners are as much concerned with the impact of their decisions on people and communities and on their quality of life, as they are with the treatment and development of space. Thus the roles, aspirations and powers of politicians, professionals, landowners and developers, organisations and community groups, and other communities of interest, are of critical importance within planning, alongside the importance of an awareness of design, and the physical organisation and sustainability of space.

2. Planning is holistic and integrative.

A key strength of planning is its ability to develop and consider the overview. A key skill of the planner is to synthesise; to recognise the core issues within multifaceted problems; and to be able to propose focused, effective courses of action, and responses to these problems. Planning is as much concerned with managing the whole environment as with the detail of any of its constituent parts.

3. Planning attempts to manage processes of change through deliberate and positive actions.

Planning is a discipline concerned with creating and coordinating action in the environment, and as such requires practitioners to be familiar with a wide range of material, with a view to taking well informed prescriptive actions in the real world of the built and natural environments. Planners are therefore, first and foremost, creative problem-solvers. Planning prescriptions require an understanding of the balances of power within societies and organisations, and the limitations that these impose upon effective planning action.

4. Planning requires appropriate administrative and legal frameworks for implementing action.

Planning invariably involves societies in developing appropriate administrative organisations and processes to regulate development within legal frameworks related to individual and collective property rights. Knowledge of such frameworks is essential for those wishing to understand planning.

5. Planning involves the allocation of limited resources.

Planning actions often result in changes in the distributions of social, economic and environmental costs and benefits on different individuals and groups within societies. Thus planning requires an evaluation of the likely impacts of decisions, and value judgements about their effects, and how they might be influenced. Planning can be used for oppressive as well as altruistic purposes, and planners need an understanding of the contexts in which each might occur.

6. Planning requires the study, understanding and application of a diverse set of multidisciplinary knowledge.

Planning requires an understanding of the relationships between underlying theory; conceptual thinking and analysis; and policy formulation, evaluation and implementation. It is an activity whose scope and legitimacy is contested, and in which a variety of justifications and views about its purposes and possible outcomes have to be understood, discussed and reviewed.

Source: Carmona *et al* (2003)

1   Planning is a means of avoiding anarchy and disorder.
2   Planning has a role in accommodating long-term shifts in social and economic structures.
3   Planning has a potential to deal with the distributional impact of change in a democratic and equitable manner.

PROBLEMS INVOLVED WITH EVALUATING LAND USE PLANNING

There are three main problems in evaluating planning. The first is that there is not always a strong linear cause-and-effect relationship between planning and environmental change. The second is the generic problem that knowledge is a contested concept. The third and biggest problem, though, is the so-called *birth control* and *displacement* effects. In essence these effects arise because those who may want to change land uses know that there is a planning system. They are therefore deterred from applying for planning permission (the birth control effect) or transfer their application from where they would like to develop land to where they think they can get planning permission (the displacement effect). It is virtually impossible to measure these effects, so we are left with the first two problems as problems that can to some extent be overcome. These are now considered in turn.

Problems with finding a cause-and-effect relationship in planning

Virtually all the plans ever produced have encountered an implementation gap between what they desire to achieve and what they actually achieve. Indeed, Pearce (1992) and Cloke (1987) have pointed out that there are a number of difficulties with the apparently simple link between policy development and implementation, and argue that in reality the link between cause and effect is blurred. In particular, Lindblom (1959) has argued that planners can never have comprehensive knowledge of the issues and thus can only muddle through by disjointed incrementalism. Pressman and Wildavsky (1973) have added that planning policies are implemented not by planners but by other people, thus inviting an implementation gap. These and other reasons for this 'implementation gap' have been aggregated by Gilg and Kelly (1996b).

The first reason is that policies are subject to change, leading to a time lag and overlapping of often contradictory policies. Second, policies can derive not just from different sections of the planning organisation making policy but also from different public or private bodies, for example policies relating to infrastructure provision. Third, policies are only guidelines and can be interpreted by decision makers in many different ways. Indeed, 'treating each case on its merits' is a hallmark of the British town and country planning system. Fourth, decision makers are nonetheless deflected from some activities by the policies in operation, and thus the real effectiveness of planning is measured by what doesn't happen. In other words the true impact of planning is impossible to measure because the world is different from what it would have been. Fifth, the evolution of policies and the lead-up to a decision on a development application are influenced more by some groups than by others. Sixth, there will be unforeseen and unintended consequences, both as the real world changes and also because of the often ignored perverse response of the free market to react in the opposite direction to a policy intention. Seventh, and finally, policy makers can only ever expect to achieve a certain success

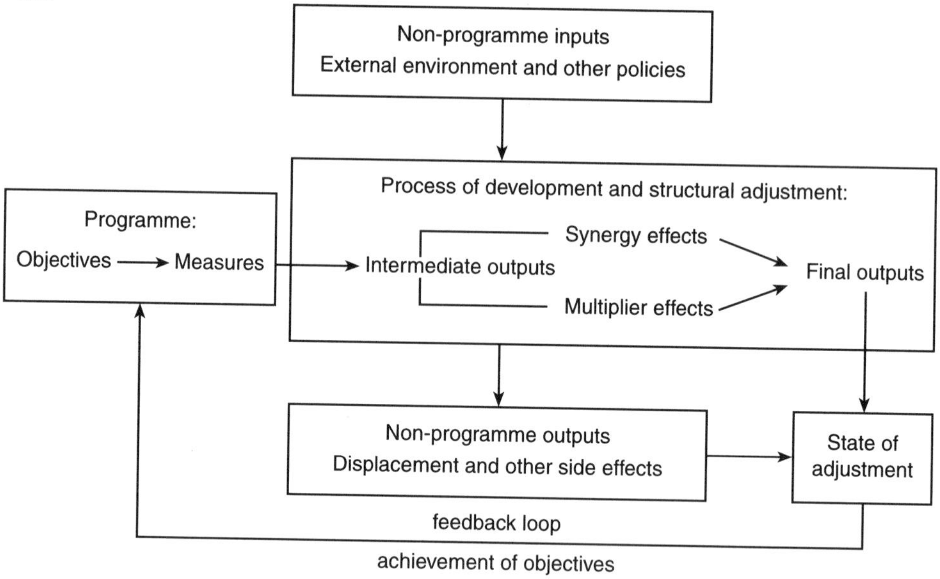

```
┌────────────────────────────────────────┐
│          Non-programme inputs           │
│   External environment and other policies│
└────────────────────────────────────────┘
                    │
                    ▼
```

```
                                   ┌──────────────────────────────────────────────────────┐
                                   │ Process of development and structural adjustment:      │
┌──────────────────────┐          │                                                        │
│     Programme:        │          │                    ┌─── Synergy effects ─┐            │
│                       │          │                    │                      ▶            │──────────────┐
│ Objectives ──▶ Measures│ ──────▶ │ ─▶ Intermediate outputs                    Final outputs│              │
│                       │          │                    │                      ▶            │              │
└──────────────────────┘          │                    └─── Multiplier effects ┘            │              │
      ▲                            └──────────────────────────────────────────────────────┘              │
      │                                                │                                                   │
      │                                                ▼                                                   ▼
      │                            ┌──────────────────────────────────┐      ┌──────────────┐
      │                            │     Non-programme outputs         │      │   State of    │
      │                            │ Displacement and other side effects│ ──▶ │  adjustment   │
      │                            └──────────────────────────────────┘      └──────────────┘
      │                                       feedback loop                          │
      └──────────────────────────────────────────────────────────────────────────────┘
                                  achievement of objectives
```

Figure 2.4   Unintended policy outcomes. *Source:* Schrader (1994)

rate. For example, Green Belt policy does not mean that no development will ever be allowed, only that some development may be allowed if there are other policies or considerations that override the Green Belt policy.

The sixth problem above, unforeseen events and the unintended effects of other policies, e.g. taxation, has been examined by Schrader's (1994) model for impact assessment and causality as shown in Figure 2.4. This demonstrates how various factors interact, and either reinforce the intended effect, e.g. multiplier effects, or reduce the intended effect, e.g. many producers react to policy signals by doing the opposite. Another problem is that the world is increasingly interconnected and, for example, if a developer is about to be refused planning permission they may threaten to take their development elsewhere or abroad. Another key problem with evaluation is the lack of detailed research and the patchy evidence. Other problems have been outlined by Corkindale (1999). First, two forms of evaluation are often used by planners and economists – planners who equate success with achieving policy goals regardless of costs; economists who look at net benefits to society as a whole. Second, there is a lack of specific objectives for policies from the government in spite of planning costing over £1 billion. Third, the complex and sectoral nature of planning goals makes evaluation very difficult.

There are thus a host of difficulties involved in analysing the impact of planning policies and the implementation gap between policy and reality. Depressingly there are also many other issues that could be discussed, for example the limited policy agenda employed by policy makers (Cloke, 1987), and the many distributional side effects that planning decisions set in train, for example on land values or social mix. Thus planners operate under many constraints, which can be graphically depicted in the six-sided 'planning constraints coffin' shown in Figure 2.5. So at this stage it is best to remember the aphorism 'To plan is human, to implement is divine.' The second problem is that knowledge is no longer seen as a value-free factor but as a commodity whose representation is fiercely contested.

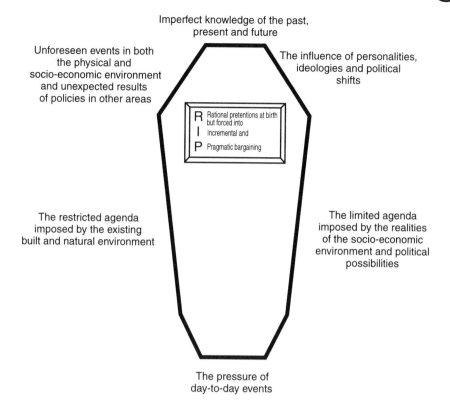

Imperfect knowledge of the past,
present and future

Unforeseen events in both
the physical and
socio-economic environment
and unexpected results
of policies in other areas

The influence of personalities,
ideologies and political
shifts

R Rational pretentions at birth
   but forced into
I Incremental and
P Pragmatic bargaining

The restricted agenda
imposed by the existing
built and natural environment

The limited agenda
imposed by the realities
of the socio-economic
environment and political
possibilities

The pressure of
day-to-day events

Figure 2.5   The planning constraints coffin. *Source*: Gilg (1996)

The contested nature of knowledge

For over two centuries since the Age of Enlightment the scientific view of the world based on empiricism and observation held sway. The related concepts of modernism with its central belief in progress and the basic goodness of human beings have also been powerful. It was felt that linking the scientific method with modernism would inevitably lead to utopia once the scientific laws linking human happiness with the physical world could be unearthed in a formula. For much of the twentieth century planners were drawn into this beguiling proposition and produced master plans depicting scientifically perfect ideal cities, for example Le Corbusier's plans for la ville radieuse.

The scientific method still remains immensely powerful and continues to alter our lives by never-ending improvements in our productive and communication capacities. But in the more complex world of human beings, and how they use the finite and fragile resources of the planet earth, serious doubts have been cast on the efficacy and neutrality of the scientific method. In particular three levels of uncertainty have been identified (O'Riordan, 1995): first, data shortages; second, deficiencies in the models used; third, the discovery that some natural systems appear to act randomly and chaotically, thus taking us beyond the knowable into a truly 'black hole'. Three main systems of formulating and evaluating knowledge have thus been debated and used since the 1970s – revised empiricism, political economy and behaviouralism/postmodernism – although a hybrid approach utilising all three has come back into vogue.

Revised empiricism

Until the 1960s empiricism was the main method employed by planners. They observed the real world and 'induced' general laws from their observations. Encouraged by the so-called Vienna school of logical positivists first set up in the 1930s, planners and other social scientists revised the empirical method by trying to seek general laws from first principles or *a priori* 'deduction'. According to O'Riordan (1995) the resulting scientific method claims that science evolves by theory building, theory testing and normative evaluation. The basic theories themselves are examined for their correctness in terms of their internal logicality, and for their consistency, that is, their inherent plausibility. These theories in turn are converted into hypotheses or propositions whose truth or applicability is subjected to analysis. Normally analysis relies upon observations of the 'real' world or modelling, through which representations of 'reality' are created to provide a more manageable basis for examination and prediction.

This approach was aided and abetted by the parallel development of systems theory and computers, which promised ways of testing general laws about human behaviour in previously impossible ways, and was enthusiastically espoused by planners. Thus the 1960s and 1970s witnessed ever more complex attempts at modelling the human environment and explaining and predicting human behaviour, which were used by planners to produce plans. The result was the classical view of the planning system as an iterative process in which plans were continually modified, as shown in Figure 2.6.

Political economy

The empirical and logical positivist systems failed to live up to their promise because they took a neutral view of the world, or in the case of classical/orthodox economics they assumed perfect knowledge by all participating players. In a reaction to these simplistic views social scientists and planners turned to the century-old writings of Marx and found there a set of theories which appeared to have major relevance to the modern world. In essence a set of workers reworked Marx's writings and developed the so-called political economy approach which focuses on the primary need for capital to accumulate and thus on the ensuing struggle between capital and other factors.

A bewildering number of sub-themes were also developed, based on the underlying structures in society caused by the imperatives of capital accumulation, for example, Althusser's 'structuralist' theory. In a connected but not synonymous development Giddens's theory of structuration attempted to find a theory for the links between structure and agency. Structure refers to the structures in society, notably organisations, while agency refers to how human actors behave as agents of change within or outside organisations. There is a long-standing debate about whether structure is more important than agency. Although these were useful theorisations they were found to be of little practical use to planners in their day-to-day work. They do, however, offer useful explanations for the impact of planning, centred on the clash between different power groups in society.

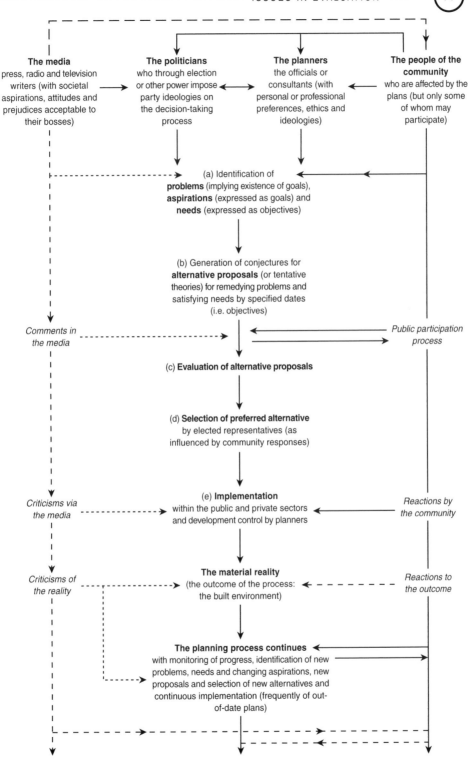

Figure 2.6 The planning process and those who may influence it. *Source*: Lichfield (1996)

Behaviouralism/postmodernism

At the same time as the political economists were reformulating Marx's general laws, another group who were also dissatisfied with the failures of the scientific method developed the idea of behaviouralism and its related field of humanism. This group, an eclectic mixture of social scientists and philosophers, argued that human beings were not essentially logical and that, especially collectively, they could act in unpredictable and illogical ways. They concluded that any endeavour involving human behaviour would need to take this into account and place humanity at the centre of any such study. In addition, language scholars, notably Derrida and the deconstruction school, argued that words are used in very different ways by different groups of people and that language can be seen not as a neutral means of communication but as contested discourse. In other words, language needed to be deconstructed in order to reveal the underlying discourses.

Thus a number of people working independently of each other produced a number of observations which have since been labelled postmodernism. Postmodernism, as the name implies, rejects the modernist project and celebrates diversity and difference, rejects the idea of meta-narratives and points out that the words used by planners to convey consensus and confidence are in fact a misuse of words to imply a scientific precision that can never exist. A key concept in both behaviouralism and postmodernism is that what really matters is not how the world actually is but how it is perceived to be by individuals, who will all have their unique interpretation. On this view planning with its normative concepts such as the public interest was untenable.

In particular, the work of Michel Foucault has become increasingly important in terms of how we perceive the world around us, notably our perceptions of power. Foucault suggests that we live in a society of socially constructed rationalities, which are shaped by discourses, constituted through power/knowledge relations and made visible in local practices. Foucault's work allows the conceptualisation of policy discourse as a complex body of values, thoughts and practices. So according to Richardson (2001) Foucault asks how, why and by whom 'truth' is attributed to particular arguments and not others. In particular, he analyses what types of thoughts, ideas, knowledges and practices become accepted, marginalised or silenced in given social conditions. This association of values and power in the construction of knowledge can be understood as the rationality of discourse. Accordingly, a set of new radical approaches to public policy has been developing which draw on discursive practices to challenge the dominant technocratic, empiricist models in policy analysis. For example, Fischer (2003) discusses how the social construction of policy problems and the dialectics of policy argumentation can be used to provide a post-empiricist analysis.

Summary

Knowledge is not the apparently certain and value-free commodity that it appears to be, and planners should make it clear at every stage of the planning process what knowledge system and assumptions they are following. Most notably, this should be done at the information gathering/analysis stage and at the monitoring/review stage. Knowledge is thus used in two different ways in planning. First, by planners who tend to show knowledge as uncontested and a series of facts in spite of the advice above. Second, by commentators who approach their evaluation from different perspectives, notably from the three major methods outlined above. Hague (2002) has provided a particularly good summary: 'Planning is a contested concept, something that is defined through social practice and takes historically and culturally specific forms' (p. 9).

REVIEW AND CHOICE OF EVALUATION METHODS

Review of evaluation techniques

Attempts to assess the impact of public policy have become very much part of the political culture. This audit or 'bottom line' (the profit or loss line in a firm's accounts) approach was pioneered by Thatcherism in the 1980s and its accountancy-driven world. It has been followed up just as enthusiastically by Tony Blair and looks set to be a major feature of public life for the foreseeable future. Indeed, all public bodies now have to set performance targets and subject themselves to regular audits. The difficulty with such approaches is that they can distort activities. For instance in the NHS, the desire to cut waiting lists led to simple operations being done ahead of more complicated ones, to the detriment of seriously ill patients. In addition they undermine professionalism with its unwritten understanding that professionals evaluate their work continually and thus do not need outside evaluation. Finally, Carmona (2003) argues that they are more often than not used as a stick to beat under-performing organisations than as a positive tool to encourage better practice.

In terms of planning, the approach is flawed since it attempts to quantify the unquantifiable – for example, the quality of life – or is based on simplistic data. For example, local authorities are set targets for the percentage of planning applications to be determined in the statutory eight weeks and plans must be reviewed every so many years. These give an indication of efficiency but not of quality. Quick decisions may be bad decisions and in planning the results are around for a long time, so why the emphasis on speed? Nonetheless, some more complex indicators have been produced. For example, 'Best Value' indicators, but these are too recent to be of much use to this book, which evaluates planning over several decades. In addition, Carmona (2003) has used evaluation techniques from management science to argue that the 'Best Value' indicators lack most of the criteria set out by such techniques. For example, the indicators do not reflect the diversity of the planning function, the indicators reflect short-term rather than long-term issues and one-off headlines, for example the speed of decision taking, have dominated rather than a more fundamental questioning of the system.

Another example is provided by quality of life indices. Approaches like this do have some value so long as one is aware of the 'health warning' that numbers cannot always convey success or failure in such a complex world as land use planning. For example, the government has produced 15 indicators for sustainable development, as shown in Table 2.1. However, it is not clear how these relate to each other and not many of them can be very influenced by planning. I have inserted the potential impact of planning on a scale of 1 to 5, with 5 being the greatest. Only two of the indicators are at P4 or P5 and even here planners are at the mercy of where developers wish to build houses. In other areas planners are only one influence among many, or in other areas – for example, traffic, air and climate – the existing infrastructure cannot be altered quickly by planning decisions.

In spite of these difficulties there are a number of ways in which public policy can be evaluated. Generically, the model developed by Gasper and George from work by Toulmin

Table 2.1 UK headline indicators for sustainable development

Indicator			Measure
Economic			
H1	(P3)	Economic output	GDP
H2	(P1)	Investment	As % of GDP
H3	(P3)	Employment	
Social			
H4	(P1)	Poverty and social exclusion	
H5	(P1)	Education	Qualifications at age nineteen
H6	(P1)	Health	Expected years of healthy life
H7	(P4)	Housing	
H8	(P2)	Crime	Violent crime and vehicles/burglary
Environment			
H9	(P2)	Climate change	Greenhouse gases
H10	(P2)	Air quality	Days of air pollution
H11	(P2)	Road traffic	
H12	(P1)	River water quality	
H13	(P2)	Wildlife	Farmland birds
H14	(P5)	Land use	% of new homes on brownfield land
H15	(P1)	Waste	

P1–P5 inserted by the author: P1 is where planning has the least likely influence, P5 where planning should be very influential.

Source: Royal Commission on Environmental Pollution (2002)

(1958) and Dunn (1994), which is shown in Figure 2.7, offers some excellent overall principles. Evaluators start with the information or data at their disposal, from which they make inferences about the policy being evaluated. These inferences are backed up by experts in their field who pontificate from their expert knowledge. Next, evaluators may call on their own or other judgemental backing, for example the moral authority of a religious text. Next, evaluators may employ analytical methods usually based on statistical tests to 'prove' a link between policy and effect. Finally, comparisons over time and/or space may be made. For example, policy X in France has produced a better result than policy Y in Germany. From these backing arguments the evaluators may then make interim conclusions but qualified by some reservation statement, for example 'it is likely that the policy is working but the impact of changes elsewhere has clouded our assessment'. In addition the evaluators will be aware that other people may rebut their conclusions, so they need to consider what counter-arguments might be made. Finally, the evaluators will make a claim concerning the efficacy of their policy.

International comparisons, as advocated above, have been made by Carmona (2003), who asked eight researchers in eight countries (including himself) to evaluate planning in their countries. Carmona suggested to his researchers that they should undertake the evaluation under three headings: first, established approaches; second, performance measurement based on processes, outputs, outcomes, impacts and indicators; third, a final evaluation based on effectiveness of approaches in delivering high-quality outcomes, key critiques of the approaches and future changes/recommendations. In a review of the evaluations Carmona concluded that, both conceptually and practically, evaluation remains problematic and requires a considerable degree of political commitment and resources before even the most basic

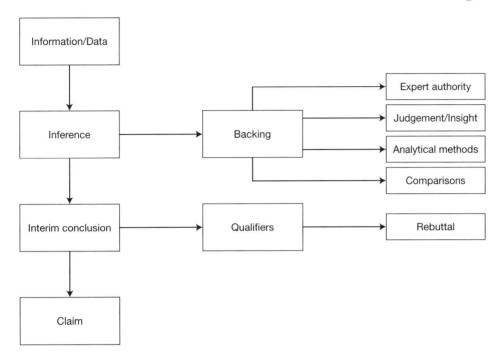

Figure 2.7 An adapted Toulmin–Dunn model for evaluating public policy. *Source*: adapted from Gasper and George (1998)

measurement processes can be put in place. However, Carmona was concentrating on measuring quality in planning and this emphasis on a contested concept may have muddied the waters.

Evaluation may also be based on a set of visions for planning as exemplified by Roberts's (2002) 'Seven Lamps of Planning':

1   *Place.* Is planning promoting effective and sustainable relationships between land uses and activities, with efficient use of the environment, with respect for the environment and above all creating places with 'identity' and 'character'?
2   *Time.* Are planning policies taking a sufficiently long-term view?
3   *Reason.* Are the planning proposals research-based, analytical and egalitarian?
4   *Opportunity.* Are the approaches flexible enough to recognise and seize opportunities without losing the overall sense of purpose and direction?
5   *Delight.* Are planners adding that extra touch of quality and care, recognising that the best for today can provide the best opportunities for tomorrow?
6   *Power.* Are planners integrated enough with power structures to achieve results without compromising ethical standards?
7   *Life.* Are planners allowing enough scope for the subtle characteristics of vitality, discretion and tolerance to let developments come alive, even, if necessary, at the expense of neatness and tidiness?

In a similar positivist vein Moseley (2003) in a major review of rural development policies and practices has argued that evaluation is a systematic analysis of performance, efficiency and impact in relation to objectives. Its purpose is not to pronounce a verdict but rather to draw lessons in order to adjust existing action and to modify future efforts. Thus the four purposes of evaluation should be, first, accountability and value for money. Second, to help managers deliver policies efficiently. Third, to learn from existing programmes. Fourth, to enhance the skills of those involved in delivering policy. The evaluation toolkit should focus on three key questions:

1   What are the objectives?
2   How far have they been realised?
3   And why or why not?

Thus Moseley advocates a focus on the chain of causation and how the chain works.

A similar method but one geared to planning has been provided by PIEDA consultants and Derek Diamond for the Department of the Environment (1992) in which the key questions were:

1   To what extent have planning policies and decisions met the needs of property developers?
2   To what extent have planning policies and decisions met the interests of conservation and the need to protect the environment?
3   To what extent has planning been effective in balancing the needs of development with the interests of conservation and environmental protection?
4   Have planning policies and decisions had the effects intended?
5   To what extent has the planning system redistributed economic and environmental costs and benefits?

The report then used Lichfield's Planning Balance Sheet to produce an Adapted Balance Sheet as shown in Figure 2.8. Difficulties arise when each box has to be filled in, since it is virtually impossible to provide quantitative evidence for each box. For example, in the 'Impacts' column how does one value amenity? One is left with qualitative judgemental evidence, albeit based on quantitative evidence where possible. This may not be a bad thing, since planning is intrinsically concerned with value judgements and should axiomatically be evaluated judgementally albeit within a clear and explicit framework.

Such a framework has been usefully provided by Morrison and Pearce (2000) as shown in Figure 2.9 in a development of the approaches outlined above. The framework also stresses the need to distinguish between outputs, outcomes and impacts. Outputs are the decisions made by the process of planning, while outcomes are the effects that planning and other factors have had on the environment, and impacts are the effects that are directly attributable to planning. In reality, it is difficult to distinguish between outcomes and impacts. However, the distinction between outputs and outcomes/impacts is sharper and so this book divides the two. First, by discussing the objectives and outputs of the planning system as a process in Chapters 1 and 3, and second, by discussing outcomes and impacts in Chapter 4 before attempting an overall evaluation in Chapter 5. The framework has been simplified and

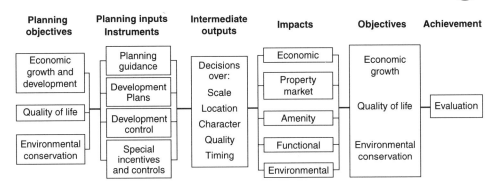

Figure 2.8 A flow chart for evaluating the planning system. *Source*: adapted from Department of the Environment (1992)

STAGE 1   OBJECTIVES

- Type of policy objective
- Nature of policy objective
- Pursuit of policy objective
- Desired standard/level of policy objective

STAGE 2   OUTPUTS

- Intermediate outputs (e.g. PPGs, Development Plans)
- Final outputs (developments control decisions)

STAGE 3   OUTCOMES AND IMPACTS

- **Intermediate outcomes** ('noise' plus planning policy)
- **Intermediate impacts** (attributable to planning policy)

- **Final outcomes** ('noise' plus planning policy)
- **Final impacts** (attributable to planning policy)

STAGE 4   THE INDICATORS PACKAGE

- Key indicators: prioritisation and selection
- Supplementary indicators
- Issues of practicability

Figure 2.9 A framework for developing indicators for land use planning. *Source*: adapted from Morrison and Pearce (2000)

modified by taking out the 'intermediate' and 'final' sub–sections and by modifying outcomes to be the physical result of a planning output and impacts to be the wider changes resulting from a planning decision. Thus an outcome might be the decision to give permission for only 60 instead of 120 houses and the impact might be an increase in house prices and difficulty for local young people in buying a house.

The framework can also be adapted to provide a simple and quick evaluation of the system or of any of its detailed components, as shown in Figure 2.10. Therefore students are advised to photocopy this figure several times and use it to evaluate the whole or individual parts of the system by taking notes as this and other texts are read by filling in the appropriate box. Students should find this particularly useful in preparing for exams, since most exam questions ask for an outline of a policy or a procedure and then an assessment of its efficacy and effects.

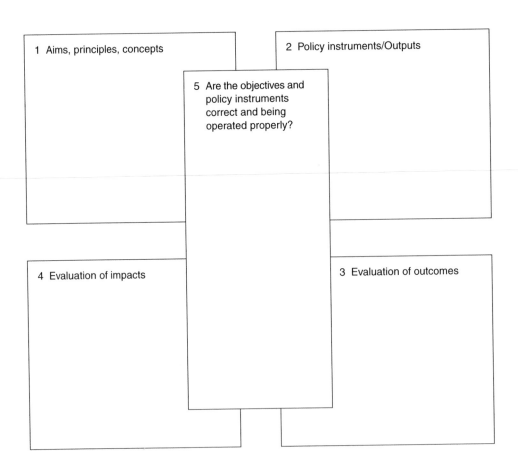

Figure 2.10   A framework for evaluating the planning system and its individual components. *Source:* author

A hybrid approach

This section has discussed some possible evaluation techniques and the difficulties that they engender. No one method appears to have a compelling logic and axiomatically the sheer complexity of planning militates against the use of only one evaluation system. Accordingly, this book uses two main approaches. First, a discussion of *outputs*, *outcomes* and *impacts* in the detailed chapters (3 and 4). Second, five summary evaluation techniques in the final chapter based on the following headings:

1   *Inferential qualitative assessments.* Because planning is based largely on qualitative judgements it can be argued that qualitative evaluations are the most appropriate, especially if based on long life time experiences.
2   *Evaluation by historical evolution.* Because planning powers have evolved incrementally due to the incremental nature of political change outlined at the beginning of this chapter it can be argued that this is the best way to evaluate planning.
3   *Evaluation by typology.* Because planning is a complex process trying to make sense and order from the human condition, typologies based on groups of activities can be a useful method of identifying patterns or trends or breaking up the process into manageable units of assessment.
4   *Evaluation by concepts and theory.* Planners are not driven by theory but there are powerful theories about power and society and how planning fits into this relationship. Accordingly, theory can provide a powerful insight into what planning has done.
5   *A hybrid approach* based on an overall evaluation of processes and objectives from a checklist that utilises all four methods outlined above is, however, the most likely to provide the broadest and most robust evaluation. This is because planning policies are delivered by planners going through a checklist of why a planning application should be given permission or not. It can thus be argued that checklists provide planners with a similar *modus operandi* and are thus the best way to evaluate the process and its outcomes.

Accordingly, I have collated all the key issues discussed in this chapter and the preface into Box 2.4(a) and Box 2.4(b), using the overall framework outlined at the outset of this chapter. The basic principle is that the argument begins with the general, proceeds to the specific and then returns to the general. In this sense the process of evaluation can be likened to an old fashioned egg-timer or the Windows icon in which information, or the sand, starts at the top as a general mass and is sifted through the waist of the funnel before regrouping at the bottom, as shown in Figure 2.11.

The evaluation methods outlined above are used extensively in the final evaluation in Chapter 5.

Box 2.4(a) Hybrid (egg-timer) approach to evaluation

1 Planning and society

To what extent is planning a mirror of its society?
Is the key role of planning to manage change with the minimum of conflict?
Whose value systems decided which issues were addressed?
Democratic versus autocratic
Public versus private
Lay versus professional
Intellect versus values
Ecology versus economy
Local versus regional

2 Planning objectives

Balance and sustainability

To balance protection of natural and built environments with the pressures of economic and social change
Twin pillars of managing urban growth while protecting landscape and heritage
Ensure sustainable land use

To set a land use framework for economic development

To ensure provision of land for jobs
To provide housing for all sections of society
To provide safe and efficient transport systems
To ensure the best use of mineral resources
Reduce deprivation and encourage regeneration
Maintaining town centres and revitalising older urban areas

To fulfil the vision of development plans

Co-ordinate investment in land and buildings
To make things happen at the right time and the right place
To ensure development works in a functional sense

Improve quality of life

Protect the architectural and cultural heritage
To ensure development is visually pleasing
Sustaining the character and diversity of the countryside
Defending Green Belts to prevent sprawl
Improving the quality of residential development

3 Outputs (plan making)

Proactive reconciliation
To what extent should planning evaluate itself through its cyclical process of plan preparation?
Are land use plans the most efficient way of promoting land use policies?
Should decision making on plans be more democratic and be transferred from expert committees?

4 Outputs (development control)

Reactive reconciliation or managing change by bargaining/consensus seeking
To what extent should development control decisions be based on land use policies?
Should development control decisions be more democratic or transferred to expert committees?
Is the twin system of plan making and development control an efficient delivery system?

5 Outcomes

Were policy responses pragmatic and incremental reactions or ideological and radical?
Has the system been effective as a system and also cost-effective for society?

Box 2.4(b)   Hybrid (egg-timer) approach to evaluation

6 Impacts

Is evaluation possible?
Has enough research been done on both the system and its impacts?

(From Objectives (item 2 in Box 2.4a))

To balance protection of natural and built environments with the pressures of economic and social change
Twin pillars of managing urban growth while protecting the landscape and heritage
Ensure sustainable land use

To set a land use framework for economic development

To ensure the provision of land for jobs
To provide housing for all sections of society
To provide safe and efficient transport systems
To ensure the best use of mineral resources
Reduce deprivation and encourage regeneration
Maintaining town centres and revitalising older urban areas

To fulfil the vision of development plans

Co-ordinate investment in land and buildings
To make things happen at the right time and the right place
To ensure development works in a functional sense

Improve the quality of life

Protect the architectural and cultural heritage
To ensure development is visually pleasing
Sustaining the character and diversity of the countryside
Defending Green Belts to prevent sprawl
Improving the quality of residential development

Wider issues

To what extent can the outcomes and impacts of planning decisions be disentangled from other decisions in
society to isolate the effect of planning on environmental and social change?
What have been the main impacts of the system, both intended and unintended?
What were the main issues that should have been addressed by the system?
Is the British system of democracy, based on incrementalism and accretionary precedent, a recipe for
disaster for modernist planning?

7 Planning and society: four objectives, three big questions and the future

To resolve conflict
To be democratic and accountable
To reconcile private and public interest
To provide equity and fairness
Are the goals of planning fundamentally different from society and thus can planning only guide and shape
change or can it make a real difference, for example by imposing greater sustainability?
More prosaically, do the delivery systems allow planning to work effectively and efficiently? In particular, is
the relationship between plan making and development control an effective one?
Do we have good enough evaluation systems and information to make a quantitative evaluation of
planning or must we rely on inferential and qualitative judgements because at the end of the day planning
is not an exact science?
Should the system be modified and if so what are the alternatives?
Should planning focus even more on being a technical exercise based on key policy criteria or should planning
break out from regulation and its narrow focus on balancing economic growth and conservation and use more of
the Gilg–Selman spectrum? And if so should the environment be accorded greater importance in planning?

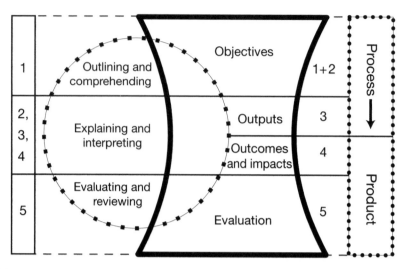

Figure 2.11   The hybrid approach to evaluating land use planning.

Main themes from the chapter and chapters to come

The key purpose of this book is to evaluate the planning system. This can be done only by situating planning within the power relations created or imposed by society. Accordingly, the chapter began by discussing the concepts of power of society within which planning operates. In particular it was argued that all societies are cauldrons of conflict partly held together by respect for the law, cultural norms and mutual self-interest. Power is exercised on our behalf and public policy sets rules for society. But using power is not a perfect exercise and there are many different ways in which it can be implemented. Planning can help to make society better by eradicating problems as society changes by providing a strategy for change. In conclusion the exercise of power over any length of time can be seen as a somewhat random sequence of events shaped by personalities and pragmatism rather than conspiratorial or dictatorial powers, with incrementalism rather than revolution ruling the day. Planning cannot be perfect because the political world it finds itself in is not perfect, but ironically planning's main claim to fame is its proclamation that it can make things better.

The chapter then considered the main objectives of planning not only as a necessary piece of description but as a crucial part of the eventual evaluation process, since planning can be evaluated only by the goals it has set up. The next section discussed the alternative types of delivery system, before the penultimate section reviewed problems in the evaluation of planning. Finally the chapter reviewed the range of evaluation techniques available. It was concluded that a series of wide-ranging evaluations using different approaches is the most appropriate, since they reflect the wide-ranging nature of planning. From these wide-ranging evaluations a hybrid or overview evaluation can be produced to address the three key questions already set out in section 7 of Box 2.4(b).

1   Are the goals of planning fundamentally different from society and thus can planning only

guide and shape change or can it make a real difference, for example by imposing greater sustainability?

2   More prosaically, do the delivery systems allow planning to work effectively and efficiently? In particular, is the relationship between plan making and development control an effective one?

3   Do we have good enough evaluation systems and information to make a quantitative evaluation of planning or must we rely on inferential and qualitative judgements because at the end of the day planning is not an exact science?

There are also issues between seeking a holistic (ideal but impossible) solution or only a partial (pragmatic) solution. The art of planning is making sufficient decisions from insufficient data, or prosaically many people see planners as quasi-legal arbitrators guiding the development/protection of the environment. This book may only be about the tools that planners have been given, their efficacy or otherwise, and how they have chosen to use them. But as Rydin (1998) and Thomas (1997) have argued planning is not just about *processes* and *outputs* but also about *outcomes* and *impacts*.

Accordingly, the book uses the three core issues of urban containment, settlement provision and landscape protection in the context of how the land use transition to residential environments has been managed by planners. This does not cover the whole gamut of planning, but it does cover the essential *raison d'être* of planning: the management of environmental change from one environment to another, notably house building.

Thus as shown in Figure 2.11, where the chapter numbers appear in the extreme left-hand column and the penultimate right-hand column, Chapter 3 discusses *processes* and *outputs* and Chapter 4 discusses *outcomes* and *impacts* before Chapter 5 provides an holistic overview.

Further reading
in addition to material highlighted in the text and in illustrative material
Some of this chapter is based on an update of chapter 1 in Andrew Gilg's (1996) book *Countryside Planning*, which provides a fuller account of planning's position in society.

Morrison and Pearce (2000) should be referred to throughout a reading of this chapter. In particular you should compare the different ways in which this book uses the words 'impacts' and 'outcomes'. Other useful introductory thoughts about generic planning issues may be obtained from:

Klosterman, R. (1985) Arguments for and against planning, *Town Planning Review*, 56, pp. 5–20.

For an excellent account of mediation as a subset of reconciling conflict see:

Healey, P., McNamara, P., Elson, M. and Doak, A. (1988) *Land Use Planning and the Mediation of Urban Change*, Cambridge, Cambridge University Press.

An international overview of public policy analysis and evaluation is provided by:

Howlett, M. and Ramesh, M. (1995) *Studying Public Policy: Policy Cycles and Policy Subsystems*, Toronto, Oxford University Press.

Web sites can be a rich source of information on planning aims but these will range from official views to some fairly wild ideas, so take care to separate the wheat from the chaff. The planning systems of other countries can also provide interesting contrasts.

3 Evaluating Planning Outputs

The chapter is divided into two main sections. First, decision making and outputs in the plan-making system are outlined, since as Chapter 1 has shown land use plans should guide development control. Accordingly development control outputs are assessed from p. 96. In each part a fairly common structure is employed, focused around the appropriateness and effectiveness of the legislation, the organisations and the decision-making processes that produce the outputs.

Learning objectives

1   To assess to what extent the legislation provides an effective framework for making planning decisions.
2   To assess to what extent the organisations provide an effective network for making planning decisions.
3   To assess to what extent the organisations have made decisions based on plan-led and other criteria.

PLAN MAKING

Legislation, organisations and processes: how appropriate and how effective?

Legislation: how appropriate?

One of the great debates in politics is about whether governments should lead the electorate or be led by the electorate and thus the extent to which legislation can change society or is conditioned by what society will allow. The post-war period has seen a fundamental shift from a society which respected authority and had grown used to controls throughout and after the war to a freedom-loving society which finds many controls irksome. At the same time there has been growing concern about environmental issues and the need to protect the countryside from development. Planning legislation has to chart a difficult course between what is in the public interest and what the public will allow.

The 1947 Act mirrored a society that had grown used to control in the war and had voted for a welfare state. It was thus appropriate for the time, but it rapidly became inappropriate as society changed rapidly in the 1960s. The 1968 Act with its emphasis on strategic plans and local plans reflected the advent of the computer and systems theory when rationality ruled. The new plans would have been appropriate but they were fundamentally undermined by the local government reforms of 1974, which were far too conservative when a radical change, to city regions, was needed. Since the 1970s legislation has modified the system in line with political ideologies and the system has become ever more divorced from the realities of social and economic forces. There is a need for a reform of the planning system and fortunately a Green Paper on planning in 2001 and a government announcement in 2002 set in motion a major reform of plan making due to commence in 2005. This new system is discussed more fully in Chapter 5.

The new system is the end result of four years of New Labour thinking and its long-term aim of modernising Britain which was initially spelt out in a 1998 White Paper (Cm 3885, 1998). Murdoch and Tewdwr-Jones (1999) and Tewdwr-Jones (1998) argue that the key features of the modernising agenda are an emphasis on regionalism and devolution, more national policy guidance, and the growing importance of a European context. However, Bishop *et al.* (2000) found that the main influence from Europe was felt indirectly from policies in other areas, for example the Habitats Directive. Apart from devolution to Scotland and Wales, where different planning systems are gradually evolving, and the growing importance of Regional Planning Guidance the agenda has yet to live up to its promise. Until it does so planning legislation will become more inappropriate, notably with regard to local authority areas which do not accord with the dynamics of daily life. The proposed regionalisation of England outlined at the end of Chapter 5 may, however, add to the alienation between planners and the planned, since few people identify with their region.

Organisations: how appropriate?

There are five main types of organisation based on the type of area involved in a hierarchy devolving from the central to the local state, namely the government, the regions, the big cities, counties and districts. These are considered in turn, but at the outset it needs to be said that there are probably too many tiers in the current system.

National organisations

At the national level, the government has failed to produce any vision of how it would like England to look in the future and there is no national plan, apart from major transport routes. The national government has, however, provided a mass of detailed policy in the form initially of circulars to local authorities and more recently PPGs. In Scotland there has been a history of national planning since the 1970s and Wales is now going down the same route. In England planning has come under several Ministries and several times there have been attempts to combine the related tasks of land use planning and transport planning in the various Departments of the Environment but this has proved to be too monolithic. After the 2001 election, planning was moved to the new Department of Transport, Local Government and the Regions (DTLR). But regional economic planning, rural affairs and protection of the built heritage were moved to the Department of Trade and Industry, the Department for Environment and Rural Affairs (DEFRA) and the Department for Culture, Media and Sport respectively. Environmental planning has thus become split up and is not well integrated, with a multiplicity of organisations – for example, the Environment Agency, the Countryside Agency, English Nature and English Heritage. Further changes continue to be made, and at the time of writing (July 2004) planning is curiously the responsibility of the Office of the Deputy Prime Minister. For the latest situation interrogate the Web.

Regional organisations

At the regional level the first organisations were the Regional Economic Planning Councils set up in the 1960s and disbanded in 1979 with the advent of Mrs Thatcher. The councils were a quango of the regional 'great and the good' drawn from most of the main regional power groups. They were appropriate in that they reflected the growing realisation that Britain was moving towards a regional society but democratically inappropriate in that they were not elected. The 1990s rebirth of regional planning was based on two administrative structures. First, regional offices of the Department of the Environment (which became the DTLR in 2001). Second, small regional planning units created from the resources of the region's local planning authorities but dominated by the counties which in an iterative process with the government's regional office produce the Regional Policy Guidance. These organisations are not elected and are thus as democratically inappropriate as their predecessors, although as Chapter 5 will show there are proposals for Regional Assemblies from 2006 onwards if regional electorates vote for them from 2004 onwards.

In a commentary on the renaissance of regional planning in the 1990s Simmons (1999) argues that two conclusions can be drawn. First, the guidance was directed at local planning authorities, which would implement the guidance in their policies, and second, the guidance was government-owned and top-down. Since 1997 New Labour has given a big boost to regionalism and devolution, especially with the setting up of Regional Development Agencies (RDAs). The RDAs were given power to allocate substantial EU funds where appropriate and to draw up economic strategies for their regions. They are thus powerful agencies in deciding where economic growth may take place and what form it may take. In common with the regional planning organisations they were not elected but appointed by Ministers from the 'great and the good' of the region, mainly leading businessmen and key figures in leading regional institutions like financial services and universities. This was endorsed by Jones *et al.* (2004), who found that key personnel already knew each other but nonetheless found difficulty in creating a new 'regional policy' culture. Nonetheless, personal relations were found to be more important than the emerging institution and so Jones *et al.* concluded that the regional state is a 'peopled organisation'.

Another issue according to Simmons (1999) is the tension between the Regional Development Agencies and Regional Planning Guidance (RPG), with the RDAs apparently able to act independently of RPG guidance. The main tension will be between the short-term nature of RDA plans and the long-term plans of RPG and the business orientation of RDAs compared with the environmental concerns of RPGs.

For the future, Simmons argues, there will be three reforms. First, their preparation will be devolved, second, the process will become more open, and third, the content of RPGs will be widened. But there will still be tensions, most notably the tension between the RPGs and the RDAs. There are also doubts about the ability of RPGs to integrate the policies of other organisations, notably transport and employment, and the danger of being pig-in-the-middle between central and local government. Simmons fears that environmental concerns will lose out to development forces and that local government will not be able to test, let alone challenge, RPG forecasts.

A preliminary study by Haughton and Counsell (2002) of the preparation of RPGs, based on 100 interviews with key participants, found that few are happy with the way their input was used. Members of the general public are particularly left out, partly because of the lack of resources given to them. Key players as in the Structure Plan process are the government's regional offices, the local authorities and the development lobby. Further work by the same authors (Counsell and Haughton, 2003) examined the tension between sustainable development and economic development strategies, discussed by Simmons above, in the context of two very different regions, the South East and the North East of England. They found some major ideological fault lines between national control over plan making and regional aspirations to devise distinctive approaches to planning for regional development. For example, the North East RPG wanted to allocate greenfield sites for employment land but the panel report argued that there was an oversupply of such land and that it would undermine national policies to regenerate inner-city areas. The authors concluded that the RPGs were a reworking of state power, with the central state only partly releasing the reins, rather than a fundamental shift to regional autonomy. For the future, though, the government has suggested the creation of elected Regional Assemblies, which may be a major step towards greater autonomy.

In a study of how the process changed after the revised PPG11 was published in 2000, based on the experience of the South West and West Midland RPGs, Marshall (2002) concluded that three changes had occurred. First, the process had been speeded up. Second, the greater emphasis on the public examination stage and the panel report had dramatically altered the balance of power and influence towards the panel and thus to the central state. This is because the panel is appointed by the state, the issues to be discussed are decided by the state, and crucially the state makes the final decision on the RPG. Third, regional planning has become a continuous process.

In common with all levels of plan in the hierarchy, compatibility is an issue, and although planning strives to achieve a balance between all the competing demands, at times choices have to be made between different demands. For example, the regional policy for the South East of England was examined by Therivel et al. (1998) under five key headings: economic, housing, retail, transport and agricultural land. Most policies were compatible except for two cases. First, the protection of the best agricultural land was not compatible with policies for the economy, housing and transport. Second, the policy to locate retail development in existing centres was not compatible with facilitating market-driven development under the economic policies.

Metropolitan organisations

In the metropolitan areas planning was thrown into turmoil by the abolition of the metropolitan authorities in the mid–1980s after their creation in 1974. This meant that these large areas of population lacked any overall strategic vision apart from the view created by the jumble of individual authorities that collectively produced the Unitary Development Plans. These plans tend to represent the lowest common denominator or what can be agreed by different factions rather than what should be agreed. The recent introduction of Mayors with real political rather than symbolic power may, however, represent a way forward and a compromise between a centralised strategic vision and the top heaviness of metropolitan authorities in charge of between two and eight million people.

County and district/local organisations

The counties and districts set up in 1974 have struggled heroically with the problems already set out and the individual changes made in the 1990s have to some extent made some more rational administrative units. But big problems remain, notably the fact that counties do not reflect the pattern of socio-economic life in most parts of Britain, apart from a few instances where remote counties are still dominated by a county town, for example, Hereford, which became a unitary authority in the 1990s. In contrast, in the 1990s Devon became a nonsensical area that does not contain Plymouth or Torbay but does contain Exeter. Most of the districts have identity problems, with names that mean little to local people, and this means that it is difficult to get the public interested in something with a title like the Wansdyke Local Plan.

Processes: how effective?

Structure planning

Since their inception Structure Plans have failed to live up to their key role of providing up-to-date strategic advice succinctly. For example, in his review of the first 10 years of Structure Plans Barrass (1979) criticised them for four failures. First, for being too broad but also too detailed and contradictory, in other words there were so many policies that one could be used to countermand another in a 'cherry picking' process. Second, they were too physical and did not pay enough attention to socio-economic matters. Third, public participation had taken too long, was too expensive and had had little impact on the resulting plan. Fourth, plans had taken too long to prepare and were out of date by the time they were completed.

Turning to forecasting, Bracken and Hume (1982) concluded that forecasting had been undertaken with little regard to planning powers, which had resulted in a confusion of wishful thinking and muddled intervention which could not give clear guidance for private and public investment. In theory, for example, local planning authorities (LPAs) and house builders should prepare joint housing studies but Rydin (1998) found these are not being pursued enthusiastically.

Most of the research into Structure Plans took place in the 1980s when they were in their infancy. Less research was done in the 1990s, from which it can be inferred that they are performing tolerably well but with the major difficulty that they are still taking too long to complete and are by and large not well fitted to their areas. An example of research is provided by 1990s work by Hull and Vigar (1998) that used 27 in-depth interviews and actor-network theory to study the preparation of two Structure Plans in Kent and Lancashire. Actor-network theory is discussed more fully from p. 97 but in essence it argues that planning is dominated by networks of actors who form an alliance of like-minded interests in order to aid decision making in their favour. They found that in Lancashire the House Builders' Federation and the county council were the key actors and that the Examination in Public was the key stage. District councils knifed each other in the back by not collectively standing up to the pro-growth lobby, so that the net result was a strong pro-development ethos in a depressed area. In Kent, Blue Circle industries created partnerships to engender an aura of growth in an area where middle-class NIMBY opposition might otherwise have prevailed.

Hull and Vigar concluded that Structure Plans are useful in managing change where growth is the issue, but that the plans are too short-term (they should have a time horizon of 30 years, not 10) and that there were key tensions between certainty, speed and democracy. Plans were also over-detailed, not flexible or general enough to adapt to changing circumstances. Thus Hull and Vigar demonstrated that the problems of the 1980s remained in the 1990s.

A study of five RPGs and 10 Structure Plans by Elson et al. (1998) found that Structure Plans followed the RPGs closely and restricted growth to a few key areas. Sustainability was viewed as balancing environmental and economic issues. However, structure planners were strongly critical of the RPGs because they were thought to have merely replicated central government advice contained in PPGs. Structure planners thought that the key role of RPGs was to set housing targets. In terms of national guidance, Baker (1995) found from a survey of 263 LPAs that central government was intervening to ensure that PPGs were followed and that restrictive policies were watered down.

From a survey of 27 Structure Plans Counsell (1999a) concluded that they had not fully embraced the concepts of sustainability, but more so in the NIMBY South than in the pro-growth North. In a commentary on five Structure Plans Counsell (1999b) found that all of them had progressed the practical application of sustainable development. However, more progress had been made in resource protection policies than in socio-economic themes where there was found to be a considerable challenge facing the implementation of sustainable development principles. Curran *et al.* (1998) from a survey of seven county councils and seven district councils found a good deal of variation in the way that plans had been subjected to environmental appraisal. The main differences were when the appraisal took place and whether it was used in the decision-making process. For the future they concluded that the introduction of Strategic Environmental Assessment (SEA) scheduled under an EU directive for 2004 will need better environmental data, the development of indicators, diffusion of good practice and training and a comprehensive system of monitoring. Carter *et al.* (2003) in a study of 40 Structure Plan appraisal reports also found that they fell short of meeting the methodological requirements of the directive, often because of the political nature of decision making.

Nonetheless Holstein (2002) feared that SEA might be seen as another technical hurdle to be overcome in the planning process rather than a means of getting sustainability to the heart of planning, as proposed by PPG1, which argues that sustainability is one of the primary aims of planning. Further research by Hales (2000) based on interviews with planning officers found that many planners viewed sustainable development merely as a bandwagon, and was only a new term for what they had in fact been doing for many years.

Bruff and Wood (2000a, b) have shown that in the major metropolitan areas sustainable development has been of secondary importance in the process of Unitary Development Plan preparation and execution. Councillors continue to be preoccupied with day-to-day issues and have left professional planners to deal with sustainability issues. This has meant that bio-diversity even in urban areas has been given attention but that more crucial issues like energy use and air and water quality have been neglected. Bruff and Wood conclude that the sustainable development aims of government policies are not being met in the major conurbations.

Local planning

Progress here has been even slower than with Structure Plans, and many areas were still without a Local Plan by 1991 when the system was changed to Area-wide Plans. Most plans were based on towns and there were very few rural plans except for pressured areas. The most comprehensive study of Local Plans remains a major piece of research by Bruton and Nicholson (1987a, b). They found that District Plans accounted for 84% of plans, followed by Subject Plans with 9% and Action Area Plans with 7%. They also found that there had been a big use of 'non-statutory plans' to fill the gap left by the lack of plans or by their being out of date.

Tables 1.2 and 1.3 have already shown the long time taken to prepare Local Plans but this cannot be blamed on an over-eager public. For example, Adams and May (1990) carried out a study of the involvement of landowners in Local Plan preparation in the 1990s. They found a high degree of involvement in Cambridge but less in Surrey Heath and the people's republic (*sic*) of Greenwich, as shown in Table 3.1, where only 57% of landowners made a representation. Most representations were made before and during the public inquiry but generally representations were not very successful except at the post-inquiry stage, but few representations were made at that stage.

Table 3.1 Landowner involvement in plan making

Involvement	Cambridge	Greenwich	Surrey Heath	Total
No. of landowners making at least one representation	33	17	18	68
Landowner representations as % of all representations	92	57	62	72

Success rates at stages	No. of representations	Success rate (%)
1 Consultation	41	16
2. Deposit and inquiry	47	30
3 Post-inquiry	7	100
4 Post-inquiry on modifications	18	11
All stages	113	25

Source: Adams and May (1992)

Elson *et al.* (1998) in a major survey of 32 rural Local Plans found that they favoured developments in towns above 10,000 people and some development in selected rural areas. Elsewhere there were strict controls in smaller villages and the open countryside with exceptions for farm-based diversification and redevelopment of major sites such as surplus Ministry of Defence sites. PPG13 was seen to have reinforced policies based on concentrating development in a few places in order to reduce travel.

Turning to the effectiveness of Local Plans, Underwood (1982) concluded that there was a good deal of uncertainty about their role and function in their early years, especially in the Thatcher years. However, MacGregor and Ross (1995) argue that the changes in the early 1990s reintroduced a pre-1980s plan-led system, and that the market might have to respond to plans to a greater extent. However, out-of-date or ambiguous plans may wreck progress to this desirable state and Healey (1988) has argued that Local Plans have been unable to provide an effective and comprehensive approach to the management of land use change.

Bruton and Nicholson (1985) point to a wider role for local planning. They argue that the broad aim of the planning process is to negotiate and agree policies, priorities and resources for change at the local level. In this process land use plans offer a means of recording the results of local agreements and negotiations. In this respect the process may be more important than the product. So in an era when planning has become 'the management of change/getting things done' the day of the plan as a neutral technical document which ignores the struggle behind its preparation may have had its day.

However, this struggle may be more apparent than real according to Tewdwr-Jones and Thomas (1998), who in their study of public involvement in plan making in the Brecon Beacons found that planners tended to lecture the public and that the debate was contained by National Park goals.

Nonetheless, Local Plans should ideally involve the public so that the implications of the fundamental nature of the planning process – the management of change through the negotiation, agreement and promotion of policies, proposals and resources for change – can be fully recognised. According to McLoughlin (1973), policies and plans of various kinds should, therefore, be seen as the outputs of a continuous planning process within the context of overall management. The weakness of statutory physical planning is its inherited emphasis on the 'plan' itself rather than the process of strategic choice, which produces it. Its overriding strength is that it forces the planning system to 'come clean', i.e. to be publicly committed about its intentions.

Summary of legislation, organisations and processes

Plan making though potentially the more creative side of planning has not attracted a great deal of interest as a process, either from academics or from the general public. This is partly because it as an abstract process not obviously related to daily life and not enough people see the connection between plans and their own lives. One way to see through the process structure is to look for patterns underneath the surface via the method of deconstruction, thus the next section provides a discussion of this intellectually more challenging approach.

Deconstructing development plans

Until recently most research into plan making has been based on empirical surveys of the process and according to Murdoch *et al.* (1999) few analysts have deconstructed plans in terms of their discursive powers. Murdoch *et al.* find this odd, since, they argue, few now accept that plans are authoritative documents but must instead be seen as documents that reflect the arena of struggle between different interests competing to determine its content. Nonetheless, plans contain a lot of 'black boxes' which are taken as read or uncontested and thus give the plan a certain irresistibility or an 'inner logic'. Murdoch *et al.* proceed to deconstruct plan making from two standpoints: the different ways in which knowledge is used and a case study of the Structure Plan process in Buckinghamshire.

They argue that there have been two views about how knowledge is used in plan making. First, there is the work of Healey (1991, 1999) which has been strongly influenced by Habermas and argues that the process should be one of 'communicative' planning in which planners seek consensus from all walks of society. (This is discussed more fully in Chapter 5.) Second, there is the Flyvberg (1998) and Tewdwr-Jones (1998) school view, which has been strongly influenced by Foucault and argues that planning is saturated by power relations. In reality the two schools are not far apart, since Healey acknowledges that there are differential power bases. The difference is one of whether the differences in power could or should be addressed by planners.

Murdoch *et al.* employ Latour's concepts of 'modalities' or 'argumentation' that divide points into either positive ones which are broadly unquestionable or negative ones which undermine a case. Another issue is the hybrid nature of planning, combining technical issues with political judgements, which, they conclude, is both a strength and a weakness.

All these concepts are used in a case study of the Buckinghamshire Structure Plan. It showed that a panel of councils and councillors set out two key 'modalities' at the outset. First, the Green Belt would be sacrosanct in the south while, second, development would be allowed in the north of the county in order to meet the housing targets and Green Belt policy prescriptions set down by higher tiers of government. These two policies were couched in terms of sustainable development and then became self-reinforcing as other policies and justifications were added, for example transport links and job prospects.

At the Examination in Public the House Builders' Federation challenged the housing allocations by questioning the validity of both housing forecasts and the Green Belt. But since these were inside the 'black box' of agreed policy at all tiers of government, the Federation was unable to break down the 'inner logic' although respondents to the researchers admitted that the allocations were political ones hiding behind technical arguments. Anti-growth groups also made little impression on the process and the inspector branded their arguments as emotional. The study concluded that the process was dominated by a small group of interlocking networks and that once the initial 'modalities' were established a pyramid of policies was constructed on a base which could never seriously be challenged.

Further analysis by Murdoch (2004) has demonstrated how at the national level a coalition between planning officials and the CPRE developed a discourse which produced PPG3 and policies aiming for 60% of housing development to be on brownfield land. However, surveys by the CPRE found that though the coalition discourse had triumphed at the national level, it could find little desire at the local planning level to increase housing densities to meet the national policies. In a similar vein, Lowe and Murdoch (2003) found that though the central CPRE attempts to play a mediating role between national and local levels of planning, county branches of the CPRE tend to go their own way. First, in areas of development pressure, like Hertfordshire, the CPRE acts as an insider group with key actors. Second, in areas with both strong and weak pressures, like Devon, the CPRE is regarded as representative of 'incoming middle classes' and local politicians actively resist CPRE's 'national' approach. Third, in areas like Northumberland where pressure is weak and large landowners dominate, the CPRE is an outsider group, perceived to represent 'southern' interests, and cannot gain access to local policy-making arenas.

In an overview of the research Murdoch and Abram (2002) come to three main conclusions about how plan making operates. First, there are the conflicting discourses of 'development' and 'environmental protection', which are potentially resolved by the balancing concept of 'sustainable development'. Second, there is the concept of governance under which the state has shifted from being the formulator and deliverer of policy to an orchestrator or conductor of networks of either 'issue networks' or 'advocacy coalitions'. Third, this develops into the concept of 'governmentality' originally formulated by Foucault, which the authors recast into:

Policy might therefore be characterised as subject to a constant struggle between, on the one hand, the construction of tightly regulated networks that permit central agencies to determine the actions of all network members and, on the other, loosely connected agencies which reshape policy in line with their own locally constructed preferences.

(p. 10).

Summary

In conclusion, planning at the national level has tended to be led by the society it finds itself in terms of the powers it has been given and the ways its processes work. But locally, plans have tended to lead society by being produced largely without the public being involved. The main problems have remained depressingly intransigent and have been summarised by Thomas (1997) as follows:

1   Outdated often before they are approved.
2   May not be compatible with other plans.
3   Plans may not contain policies for new types of development.
4   There may be arguments about the interpretation of policies.
5   Confusion between land use aims and social economic aims.

In particular, consider if a plan drawn up on the back of an envelope by a pub-based focus group might be more effective in relation to Kitchen's law, which Kitchen (1996) after a lifetime as a planning officer proposed as follows. Time over which the plan is useful divided by the time taken to prepare the plan should be greater than 1.0. In other words we should get at least as much value out of plans as the effort put into them, and a plan should be in force longer than the time taken to prepare it.

Discussion

The main theme is that planners have massively changed their role from producing plans that they would implement to producing plans that other people will implement. Planners have thus sought to work with other organisations to provide plans that will be acceptable to others. This is different from a view in the 1970s that was perhaps coloured too much by plans that tried to prevent change or were blighted by the economic crises of that decade. In this move to positive planning, in which development should be allowed unless there are clear-cut reasons for refusal, planning departments have often lost their title and been renamed with words like 'development' or 'property services'. This reflects a changing role of planning to facilitating or managing change, although this may begin to lose sight of the balancing role and the conservation role set out in Chapter 2.

A second theme has been an emphasis on speed and efficiency since the 1980s now encapsulated by Best Value systems and in the 1990s by the Citizens' Charter. All these stress that planning is a public service, which should make speedy, consistent and transparent decisions. Unfortunately, plan making has been slow, not very consistent and only thinly transparent. Accordingly, current proposals, discussed in Chapter 5, recommend that some plans should be abandoned to be replaced by Regional Plans and at the local level plans will be produced only where major development has been ordained by Regional Plans.

A third theme is that plan making has been changed three times but that none of the changes have been very effective, partly because the administrative units have been faulty. Related to this problem is the hierarchy of plans, with three levels now existing. Because of the overlapping time horizons of these plans there has been a good deal of uncertainty about which plan has precedence over another or for how long it will be extant. As Chapter 2 has emphasised legislation in Britain is incremental and cautious and this is emphasised in Box 3.1

where it is recognised that many of the problems of plan making may have only marginal solutions. Nonetheless, the government made fundamental proposals for reforming plan making in 2002 and 2003 which will be discussed more fully in Chapter 5.

Box 3.1 Development Plans: common criticisms and reform potential

Fundamental problems which can be tackled only at the margins

1 Slow in production, notwithstanding the efforts of the planners working on them to move through each stage as quickly as possible.
2 Can date very quickly, although not all elements decay at the same rate.
3 Subject to the rapid rate at which government policy can change, for example through the PPG system, often within the lifetime of the plan preparation process.
4 Arguably too biased towards the interests of objectors – local plan inquiries are inquiries into objections, not into plans.

Some change possible but change will bring other problems in its wake

1 Ambiguous in relation to individual cases, perhaps because, in their tendency to comprehensiveness, they contain rafts of policies which are perfectly sensible in their own right but which conflict with each other in the guidance they give at the site-specific level.
2 Better at broad policy than site-specific detail. Site-specific proposals often leave little room for interpretation and flexibility, whereas broadly worded policies not only imply ample scope for manoeuvre but also are often deliberately drafted to provide it.

Probable worthwhile gains to be made

1 Difficult to understand, as they are not usually written in customer-friendly ways. Thus make sure that new plans are written in simple, jargon-free English.

Source: after Kitchen (1998)

In conclusion, plan making has not been as rational as axiomatically it implies that it should be. This is largely because, as Cherry (1979) has so succinctly put it, 'Planning legislation is the story of the incremental adoption of measures imperfectly conceived in respect of problems only partially understood' (p. 316).

Thus planning suffers from two problems. First, we do not understand the world around us, and second, as a society we are not willing to tackle the problems that we do perceive with enough rigour and vigour, and when we do we do it incrementally. This is because planning is a prisoner of the society it finds itself in and can only reflect rather than change that society. In this sense planning is fatally flawed since its claims to make the world better are constrained by what society is willing to let it do. This is vividly displayed by car driving. Most of us can now drive and afford a car, we cherish the freedom it gives but curse the traffic congestion. Collectively we know that planners should do something about it but no political party has the courage or stupidity to offer a radical alternative to our car-based society. We want the ends but not the means.

The challenge for plan making is to produce a vision of how we can have a car-based society that marries the virtues of community and cityscape exemplified by cities like Oxford and Edinburgh with the mobility demanded by the modern economy and lifestyles based on recreation. Sadly, our history of plan making has shown that it has been rooted in short-term thinking and solving current problems rather than offering an alternative to sporadic development.

DEVELOPMENT CONTROL

Development control is important since about 10% of Britain's gross national product (GNP) is accounted for by investment in land, buildings and works (Thomas, 1997). The main purpose of this section is to evaluate development control in terms of how it is set up as an administrative process, via an evaluation first, of its legislative and organisational framework and second, of the decision-making processes involved in development control.

Learning objectives

1   To assess to what extent legislation has provided development control with an effective range of powers and options.
2   To assess how appropriate development control is in managing change as part of the planning system, notably with regard to implementing development plans.
3   To assess the effectiveness of the process in reaching consistent and predictable decisions in a reasonable time in the general context of the 'public interest'.

Four key points about development control

At the outset it is important to reiterate the four key points made about development control in Chapter 1

1   It has stood the test of time and has remained essentially unaltered except for detailed changes since its inception in 1947. This must say something about its quality.
2   It is the 'sharp end'/'coal face' of planning. It does not have the glamour of plan making and does not tend to attract 'high-flyers', and traditionally the job was taught to people training to be planners from school, rather than graduates.
3   It is where the impact of planning is felt and can be measured in terms of what has and what hasn't been allowed, especially with regard to the effect of the system on the geography of Britain.
4   But like birth control its effects are hard to assess, because what hasn't happened can't be known. This can be called the 'deterrent effect' in that planning applications are not made if there isn't a reasonable chance of success. In addition there is the 'displacement' effect which has affected behaviour by displacing planning applications elsewhere. The 'deterrent' and 'displacement' effects thus make measuring the effects of development control a methodological nightmare.

Legislation, organisations and processes: how appropriate and how effective?

Legislation: how appropriate?

The first point to make is that development control remains in essence the same as when it was first set up in 1947. This is even more striking when it is recalled that development control was meant to play only a residual role, but has in fact become the main instrument by which planning is delivered. Therefore it can be argued that the legislation has been very appropriate. However, this is partly due to the flexibility that the legislation gave in allowing the GPDO and the UCO to be modified to reflect changing circumstances and political ideologies.

Organisations: how appropriate?

The organisational structure of development control is, however, less satisfactory and may be biased in favour of certain interests, nationally and locally. It is tempting to think of planners as if they were all-powerful and all rational, but in reality they have to work not only in the context of the legislation but in the context of national and local power relations that severely constrain their decision making. Planning commentators used to report on development control as if it was a straightforward exercise in rationality. But from the 1980s commentators and researchers have increasingly used the concept that society is structured by key organisations and thus that actors work in networks between and within these organisations in the context of their working relationships.

The concept of actor-network relations in development control

This concept recognises that human beings have to work within the structures of society. Human 'actors' therefore act in the context they find themselves in but also in the wider context they form of networks of contacts. However, other 'actor networks' may have different opinions and so a struggle of interests may ensue between different actor networks. Some people have tried to divide these approaches into a schism between agency and structure in which either structure or human agency dominates. But a new school of thought emerged throughout the 1980s and 1990s that argued for the two concepts to be united in hybrid concepts. Thus planning researchers developed three broad concepts of how planning officers worked:

1 *Actor in context*. Planners are humans with prejudices but these are to some extent subsumed by their professional training and the context of the planning office and their wider relationship to society, notably the priority given to capital accumulation.
2 *Actor networks*. People like human contact, especially with like-minded people, so actors form networks. In addition society and legislation impose informal and formal networks on people. Planning can be visualised as a series of structured interactions between actors in an actor network
3 *Struggle*. The development control system has a well established actor network but not all members of the network will agree all the time. Thus an element of conflict and struggle is implicit in the system. However, other networks may have deeper levels of disagreement with the main network and there is a struggle between insider and outsider networks. In summary, development control can be seen as a struggle for power between different actor networks by people who work in the cultural context of their own professional organisation.

Studies of actor-network relations in development control

This approach was pioneered by Short *et al.* (1986, 1987) in the mid–1980s. As shown in Figure 3.1, they studied the relationship between the central state, house builders, local planning authorities and pressure groups. Traditionally, this matrix of interaction had been thought of as neutral, with each group having equal power, but Short *et al.* found that there were significant differences between and within the four key groups.

First, there was a clear tendency for large-scale negotiators to be the most successful, whilst naive and aggressive approaches were the least successful. Not surprisingly, most volume housebuilders employ full-time staff to develop good relations with local planning authorities throughout all stages of the planning process in order to build up 'banks' of land with planning permission. The process thus favours big builders at the expense of small builders or individuals.

Short *et al.* (1987) also found that the public are unequal in terms of their involvement with the planning process. First, 68% of members of pressure groups come from owner–occupied houses and are thus representative of the 65% of Berkshire people who live in their own houses. But the 22% of people who lived in council houses accounted for only 5% of members of pressure groups. Surprisingly, 26% of members came from other types of accommodation. However, these people were mainly 'getters', i.e. people who wanted planning to get development going, rather than 'stoppers', people who wanted to stop development. The vast majority of stoppers came from house owners, but a degree of ambivalence among house owners was shown by reasonably high proportions of 'getters' also coming from this tenure type. This probably reflects a schism within the middle classes between those who work in the private sector and can be deemed to be 'getters' and those who work in the public sector and may be deemed to be 'stoppers'. In contrast council tenants were either neutral or in favour of development.

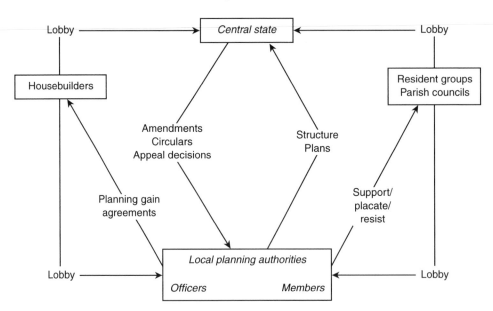

Figure 3.1   Agent interactions in the planning process. *Source:* Short *et al.* (1987)

In terms of the type of activity engaged in, commenting on individual applications was the most frequent activity. Plan making was much less frequently engaged with. This reflects a basic problem that people tend to be involved with planning only when it affects them personally and that more abstract activities like plan making do not attract so much interest. This is particularly marked in the lower engagement with the more remote Structure Plans than closer-to-home Local Plans. In conclusion Short *et al.* (1986, 1987) theorised the following process:

1 Anti-growth opinion of locals is expressed by refusals by the local state.
2 Appeals to the central state by builders are thus frequent.
3 The central state justifies upholding appeals and allowing higher growth by adding 'planning gain'.
4 At the plan-making level the central state lifted the housing provision from 31,000 to 39,600 houses after the EIP of the Structure Plan.

Short and his team thus concluded, using the language of political economy fashionable at the time, that planning was a tool of capitalism. In more detail, planning was merely 'legitimising capital accumulation' by giving the impression of local democracy and power over development while central government reserved the right to restore the imperative of capital if it was threatened. This was exacerbated by the Tory government of the 1980s, which favoured development.

This view was endorsed by Pacione (1990, 1991) from a similar study of planning outside Glasgow. From this work Pacione concluded that planning decisions highlight the way in which the state adjudicates between divergent interests and achieves its dual imperatives of capital accumulation and socio-political legitimation. In other words planning is a cosmetic exercise which deludes people into thinking they have a say, whereas in reality big business rules the agenda, albeit implicitly.

Similar findings were reported by Bedford *et al.* (2002) in their study of two planning applications in London. They interviewed not just the key actors but, notably, the 15–20 local residents who had objected to each case. They found that in spite of much emphasis on new participatory practices the residents felt disempowered by a process that consults more than it did but still does not permit those outside key networks to gain effective purchase on the decision-making process. The authors concluded that:

> As long as the structures and practices associated with the development control process continue to constitute and reproduce power relations that privilege property owners and powerful economic interests, public participation, no matter how wide, is unlikely to enhance confidence in public institutions or empower citizens
>
> (p. 329)

A more humanistic interpretation was provided by Rydin (1985) in her work on residential development and planning in two contrasting areas of Essex, Epping Forest, a leafy green area of upper middle-class housing, and Colchester, a town of mainly skilled working-class people. She used interviews with planners, a study of plans, development control data and a survey of applicants to produce two key findings. First, permission rates varied widely by both area and the type of development. (Similar studies will be analysed more fully in Chapter 4). Second, the reason for this variation was provided by the ethos of people in each area who together provided a powerful network of agreed interests. In Epping Forest the ethos was one of high house prices reinforced by the extra scarcity provided by strict planning controls. In Colchester work was the main ethos and so development was not only accepted but welcomed. Thus planning is not always a struggle.

However, in much of rural and suburban Britain a massive anti-development ethos is building up, commonly termed NIMBYism (Not In My Back Yard). These people hold the view that, having gained their 'place in the sun', i.e. a nice house in a pleasant environment, they want to keep it that way and so lobby for anti-development planning or as Buller and Hoggart (1986) eloquently put it non-decision making. In other words certain unpalatable items like the demand for housing are kept off the planning agenda, thus avoiding a decision which may be disliked by the NIMBY lobby. Such groups use the planning system not only to protect their asset but also in a virtuous circle for them: restrictive planning actually appreciates their asset, since it makes it scarcer and thus more desirable. Murdoch and Marsden (1994) have examined the land development process in Buckinghamshire using this concept of middle-class self-interest. The main policies in the area are shown in Figure 3.2 from which it can be seen that landscape restraint policies dominate, especially in the south of the county. Settlement growth is restricted to very few areas, although in a classic planning compromise one of these areas, High Wycombe, is in the middle of the AONB.

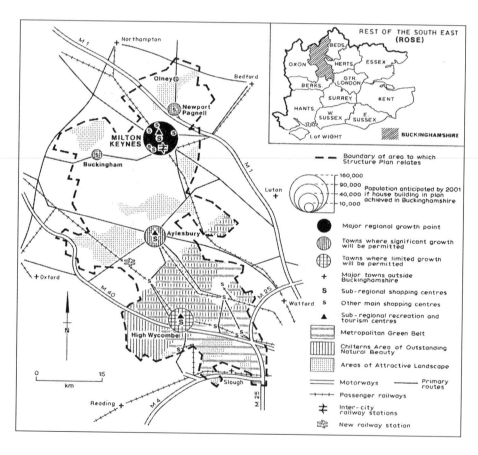

Figure 3.2   Development Plan policies in Buckinghamshire. *Source:* Murdoch and Marsden (1994)

Murdoch and Marsden demonstrated how the middle classes in the area have created a vision of 'traditional village communities' to create a strong anti-development stance among village residents, and thus have constructed an image of exclusivity which is mutually reinforcing, particularly in the way in which the planning system operates. This has prevented farmers from diversifying into certain activities, and has funnelled them into activities which, for example, reuse old buildings in middle-class ways, or develop land into socially acceptable golf courses.

Murdoch and Marsden argue that the middle classes have not used the real arguments to advance their case, namely the desire to exclude less well-off people or people who do not conform to being white, English and family-oriented. Instead they have used arguments based on 'acceptable' concepts of peace and quiet, and landscape preservation. Woods (1997) in a study of house building in Somerset has confirmed Murdoch and Marsden's view of planning being used to create middle-class villages.

The work by Murdoch and Marsden was part of a very long research project by a larger team into rural change which was aggregated into a book in 2003 (Murdoch *et al.* 2003). This work had begun with a Marxist orientation but as it progressed it produced theorisations more closely related to actor-network and conventions theory, with two main assumptions. First, top-down policies are no longer received passively but are mediated locally. Second, local action leads to local differentiation. Collectively, four main parameters of activity – economic, social, political, cultural – have combined to produce four types of countryside: the preserved, the contested, the paternalist and the clientelist.

In more detail conventions theory proceeds from the assumption that any form of co-ordination in economic, political and social life requires agreement of some kind between participants. Such agreement entails the building up of common perceptions of the structural context. These become conventions. Spaces are therefore marked off from one another by practices of consensus building within networks and by economic, political and social conflicts between networks. In other words the countryside emerges from processes of co-operation and competition and these processes give rise to an undulating geography of networks and conventions.

Networks also overlap at the international, national and local levels. Interest therefore shifts to how decisions taken in one area are transmitted to actions in another. At the local level networks tend to cluster around shared infrastructures, local labour skills, economies of scale and by sharing skills and knowledge. Nonetheless, personal preference can still have a big say in investment decisions. Two types of village emerge. First, the traditional one based on kinship, spatial proximity, close social co-operation and an agricultural orientation. Second, counter-urbanisation areas where networks come together around shared interests, notably a desire for community life and a good environment.

Murdoch *et al.* then use these theorisations to examine the outcomes of the 1947 Act. The Act was based on two core beliefs: maximise food production and protect farmland from development. But this has been undermined for three reasons. First, agriculture has protected neither the rural economy nor the rural environment. Second, new economic uses of rural space have emerged associated with mobile jobs. Third, villages and protected landscapes have become desirable places to live. Regional differences in these three factors have, however, demarcated contrasting environments of action and ensured that different rural areas are placed on different trajectories of development. Thus three types of rural area in terms of planning actor networks emerge: the preserved countryside, the contested countryside and the paternalistic countryside. Three case studies of each are discussed.

In the preserved countryside (Buckinghamshire) an anti-development network and conventions at the county level divide the county into two, with the south being dominated by preservation networks. At the local level the southern protected area is reinforced by anti-growth conventions but in the growth areas there is a clash with conventions based on protecting neighbourhoods, community and the environment with conventions of growth, need and market demand. Overall the local anti-growth sentiment has begun to work upwards into the regional level, leading to a preserved countryside not just in parts of Buckinghamshire but across more and more of the South East.

In the contested countryside (Devon) there are two groups and areas. First, development networks of traditional residents, mainly farmers and small businesses. Second, environmental networks of middle-class incomers. Because neither group can gain the upper hand the countryside and planning policies are contested. Finally, in the paternalistic countryside (Northumberland) the owners of large landed estates have no interest in seeing their land developed and the planning system is marginalised

Summary of actor-network studies

These studies have emphasised the wider context in which planning decisions are made and have queried whether development control is merely a tool of wider structures. These structures, however, contain two contrasting forces: first, the interests of business seeking to develop land, but second, the political power of middle-class house owners seeking to preserve their property and environment. The studies above have suggested that big business has not always been successful and that planning can be a genuine power struggle, albeit between some players who are more powerful than others. It is now time to examine some other issues, notably national figures on decisions, before coming to some conclusions about the efficacy of development control as a decision-making activity.

Processes: how effective?

Development control decisions

At the outset it is important to note three key facts about the decisions in order to set the subsequent arguments in context.

The first fact is that there are about 450,000 applications a year, of which about 350,000 a year are for physical development. About 10–20% are refused, about 25% of which are appealed against, with an approximate 70% failure rate. The overall figures vary with the economic cycle and increase with house price booms, as for example in the early 1970s and late 1980s, when the number of applications rose rapidly to over 500,000 as house builders tried to cash in on the boom. Another factor influencing the rate of permissions can be political. For example, in the 1980s, encouraged by Mrs Thatcher's deregulation stance, the number of appeals rose to 30% and their success rate rose to 40%. This period has been dubbed the 'appeal-led' era. But when more restrictive planning was forced back on to the Conservative Party agenda in the late 1980s by a middle-class backlash from the leafy green shires and Green Belt areas the appeal rate fell back to the long-term norm. Similarly there was a return to the long-term average of 500,000 applications a year in the 1990s after the housing boom of the late 1980s.

A second fact is that the Planning Inspectorate by and large upholds LPA decisions, thus suggesting they are operating the process appropriately. Appeal decisions have been examined by Tewdwr-Jones (1994) in the South Hams. He began by demonstrating that the South Hams were reasonably typical of England and Wales, except for appeals allowed, where the South Hams figure was 26%, compared with the then national figure of 33%. Next, he analysed the effect of landscape designation and found that out of the total of 134 appeals in the area, 112 were in a designated area (although this reflects some double counting, as two designations may overlap). However, the success rate varied from 14% to 21%, nearly always less than the overall success rate of 26%. The exception was the coastal preservation area, where 50% were allowed but only 10 appeals out of 134 in the total area were made.

Turning to the reasons for decisions both the South Hams and the inspectorate personnel used very similar policies from the Devon Structure Plan. Three policy areas dominated: access and traffic issues; the protection of designated landscapes; building inside designated landscapes and not in other villages or the open countryside. Finally, Tewdwr-Jones compared the use of Local Plan policies. He demonstrated once again that the two organisations used very similar policy reasons. This time, however, the types of policy were different, with detailed issues, notably the type of boundary to be used, being the most frequent policy consideration. Nonetheless, local policies on countryside protection are also important. Tewdwr-Jones could conclude that both strategic and local planning policies were standing up well on appeal.

Wood et al. (1998) and Punter and Bell (1999) have shown policy was playing a greater part in deciding appeals in the late 1990s. Wood (2000) has demonstrated that as the approval rate rises the appeal rate falls, ranging from authorities which turned down up to 25% of applications to some who turned down only 5%. However, when Wood plotted the rate of appeal allowed on the same chart no relationship was seen between the different rate of appeal success and the rate of planning approvals. In seeking an explanation Wood did a chi-square test comparing appeal approval rates in four different areas: Cities; Large towns; Town and County; Rural. He found that rural districts were far more likely to be supported by the Planning Inspectorate than urban districts. Wood constructed a new correlogram based on appeals allowed and the appeal rate, as shown in Figure 3.3, which allowed him to hypothesise four types of local planning authority based on the pressure for development and the support given to policy by the inspectorate:

1   *Area A.* Low pressure/high support (mainly cities and large towns).
2   *Area B.* Low pressure/low support (mixed urban–rural).
3   *Area C.* High pressure/high support (mixed urban–rural).
4   *Area D.* High pressure/low support (mainly rural).

Only areas A and D appear to have a strong urban–rural dimension and so Wood also plotted out two maps. The first map of appeal rate by district showed that appeals are far more common in the South East than anywhere else, reflecting the severe demand for development in this dynamic region. Second, as shown in Figure 3.4, Wood plotted the pattern of A, B, C and D areas. Low pressure/low support areas predominate in the north Midlands and Lancashire and Yorkshire. Low pressure/high support areas are found in East Anglia and the North. High pressure/low support areas are found in the South East and suggest that in the 1990s the central state is still operating in the same way as suggested by Short et al. (1987) in the 1980s, by overturning NIMBY resistance to development at the local authority level. Finally, high pressure/high support areas are found in the attractive rural areas of the South West and the AONBs of the Cotswolds, Chilterns and Downs of south-east England. This

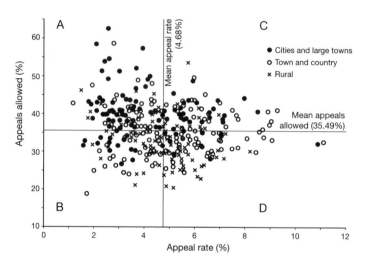

Figure 3.3 Appeal rate, by type of local authority. *Source*: adapted from Wood (2000)

indicates that the pressure for development as found by Brotherton in the 1980s (See Chapter 4) has carried on into the 1990s. However, as before the pressure is being resisted at both the local and the central state level.

This work suggests strongly that local decisions are being endorsed by national decisions and that both levels of decision making are consistent with each other. From a major study of the appeal process which used the Planning Inspectorate's database Punter and Bell (1998) found that the refusal rate did not vary much across the country except for the older industrial and mining areas where the rate was markedly lower. The appeal rate was also even except for the case of urban centres and inner London. Finally, the dismissal rate was much the same everywhere. These figures show that the appeal system does not have any significant geographical variation and that local planning authorities are presumably applying policies equally.

In an empirical analysis of appeal decisions Bingham (2001) concludes that they are more plan-led than aggregate statistics suggest. There are, though, frequent disagreements between decision makers over the application or interpretation of plan policies. This strongly suggests that flexibility and discretion remain important features of development control at both the local and the appeal level.

A third fact is that since the 1980s speed has been emphasised, with a target of most decisions being made within two months. The government has set up regular league tables in which slow LPAs are named and shamed. This is not really appropriate, since like all league tables of public service performance targets it ignores the raw material and circumstances of the area. Furthermore, as the influential Audit Commission (1992) noted, speed is not an appropriate criterion when buildings may be around for decades if not centuries. Quick decisions may not be good decisions and the process is not being operated appropriately in this regard.

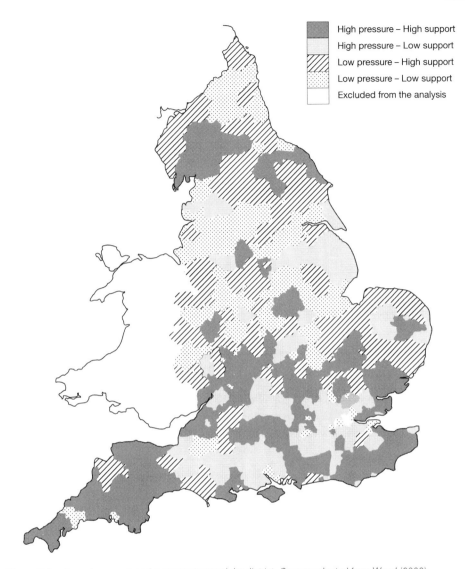

Figure 3.4 Appeal pressure and support at appeal, by district. *Source:* adapted from Wood (2000)

How decisions are made

The first issue is that the process is reactive. It can work well when it can react to heavy development pressure by 'cherry picking' the best developments. For example, several supermarkets may want to locate a new store and the LPA can use conditions to provide the best 'planning gain' from one of the applicants – for example, a new leisure centre or paying for a new access road. It does not work so well when it is an area of decline, since axiomatically there are few planning applications to decide on. It can signal via plan making and by giving quick and flexible permission that development is welcome but if the private sector is still not interested there is little it can do.

These problems are manifested not just in inner-city areas but also in the countryside. For example, Rowswell's (1989) study of planning in East Sussex between 1971 and 1981 found surprisingly that growth in the key villages (i.e. those selected for growth) of 9.8% was less than growth in the non-key villages of 12.3%. Rowswell attributed the differences not to planning policies but to the different attitudes of landowners to development, with some landowners failing to cash in their land for development even in key villages, frustrating the ambitions of both planners and developers. Spencer (1997) partly confirmed this thesis but stressed the interconnectivity of landowners/planners and other actors in the rural development process.

In contrast Gilg and Kelly (1997b) found that farmers used all sorts of covert arguments unrelated to farming to gain planning permission in north Devon. When these failed they adopted a range of behavioural tactics, which undermined both central and local state policies and cut across all sorts of other divisions like social class or political party. For example, insisting on a site meeting at which only those committed to the application turned up. In so doing they managed to gain planning permission in many cases against the professional recommendations of the planning officers and the rest of the planning committee. This led to over a quarter of decisions over agricultural dwellings going against the recommendations of officers.

Cloke (1996) has reported on similar findings in Wales, where discrepancies between officer recommendations and committee decisions seem to be at their highest in the more rural areas. Likewise in England, in one of its most rural counties, Cornwall, a senior civil servant had to be called in to investigate decisions that seemed to over-favour the friends of planning councillors. Thus it seems that councillors can operate the system inappropriately, if legally. As we have already noted, graft and corruption are fortunately rare, although they do occur. For example, in March 2002 two councillors in Doncaster were jailed for exerting undue influence in obtaining planning permission for a housing estate against the policies in the Local Plan and the recommendation of the planning officer.

Planning officers

Planning officers are only human and although like all professional people they are trained to keep their private views out of their decision making it is inevitable that from time to time this will break down. Indeed, in a study of planning gain based on development control data and interviews with key actors Bunnell (1995) found that the behavioural attitude of planning officers was a key variable. This was because two different chief planning officers operated the process quite differently, based on their personal beliefs about how planning should work. He concluded that variations in local authority practice reflected the varied professional beliefs, attitudes, philosophies, styles and temperaments of individual chief planning officers. He used this finding to argue that because most of the literature on planning decisions seems to assume that local planning authorities behave largely alike, this has blinded researchers to the importance of trying to identify and analyse factors and conditions which shape local planning authority behaviour.

In a theoretically advanced analysis Tait and Campbell (2000) examined the detail of decision making in a town-centre development scheme. Analysis techniques based on the communicative theory of the German philosopher Habermas (explained more fully in Chapter 5 in the analysis by Healey) revealed the different types of knowledge used. Planners used technical knowledge that was quantifiable while councillors used local knowledge based

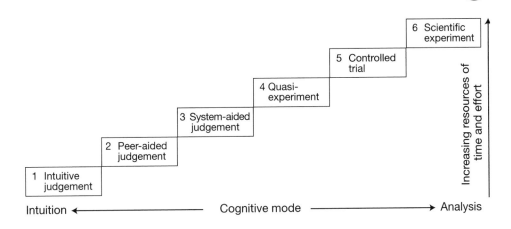

Figure 3.5 The cognitive continuum of decision making. *Source*: adapted from Willis and Powe (1995)

on experience. Further analysis, this time based on the work of the philosopher Foucault, used discourse analysis to identify professionalism and democracy as the difference. In both cases there was a gulf between the thinking of the planners and the councillors.

However, planners do not have the time, resources or energy to acquire perfect information on every application and may well rely on instinct and judgement. Indeed, Willis and Powe (1995) as shown in Figure 3.5 have argued that most planners go only one or two steps up the potential six-step cognitive continuum, through lack of either time or inclination. Most decisions, they argue, are based on intuitive judgement or peer-aided judgement, then policy issues are added in to justify the decision. However, this may not be too problematic in that planning is a political process of choosing what is in the public interest and in this regard relying on personal as much as professional evaluation may not be so bad.

Furthermore, Kitchen, writing from a lifetime of experience as a planning officer in Manchester, has argued that only about 10% of applications are the subject of real negotiation. Another 10% can be improved by straightforward negotiation but of most applications around 60% are straightforward and are delegated to officers. This leaves only 10% of applications which cannot be improved enough by officers and so they have to be refused.

In an attempt to provide a conceptual framework of the arena in which planners operate Campbell and Marshall (2000) have argued that the normative and rational decision-making model has increasingly been questioned both theoretically by postmodernists and politically by the New Right. Thus the hegemonic status of planner as a professional acting in a value-free vacuum has increasingly been questioned. Campbell and Marshall then adapt Bolan's 11 moral communities of obligation into 5 categories of obligation: individual values; the profession; the employing organisation; elected members; the public.

Individual values are derived from life experiences, and the planning profession is deemed to be young, white and male. The planning profession is state-orientated and split between vague concepts of professional specialism and universal generalist. The employing organisation used to be cosy and compartmentalised, but since the 1980s managerialism, performance indicators, corporatism and partnership (entrepreneurial consensus) have transformed the working environment of planners. The net result is that an audit and contract culture has led to financial considerations conflicting with land use considerations. The tyranny of league tables may also lead to less negotiation and more black-and-white decisions. The harmonious balance between collective welfare through procedural justice may have been disrupted by the move towards corporatism and economic efficiency.

Elected members can spectacularly collide with officers where officer advice is rejected. In that case planners must yield to the wishes of their political masters and may have to go against their professional judgement. The public are not monolithic and there is no easy answer to assessing public interest. The audit culture favours applicants, since this is where performance is mainly measured.

Campbell and Marshall conclude that there is a big need to measure the public interest so that planners can better gauge success in achieving the public interest. There are two main approaches. First, utilitarianism, the greatest good of the greatest number, possibly measured by the best market conditions. Second, the need to compensate for inequalities and where individuals are mistaken about what they think is in their best interest. The authors argue for the rehabilitation of this second approach, based on the common good and collective responsibilities.

Summary

In conclusion, all the studies above have shown that there is only one potential big problem with the process, an implementation gap between policy and development control decisions if the process is not operated properly. It is not clear, however, where the blame for the implementation gap lies. Some studies argue that it is the unequal distribution of power between the actors, some studies argue that behavioural factors account for the difference, but few studies claim that the system itself is at fault, only the way it is operated. Indeed, we finish this chapter with the key point that development control has stood the test of time. Indeed, there has been only one major review of its work by the government, and that reflected the issue of speed with which the mechanism was operated rather than faults with the mechanism itself.

The Dobry report

The Dobry report (Dobry, 1975) was commissioned by the government after the house price boom of the early 1970s had led to applications jumping from 400,000 to 600,000 a year. Inevitably delays were caused and the government called on George Dobry, an eminent planning QC, to hold an inquiry. His main finding was that the mechanism was fine, but that it was being operated inefficiently. He therefore proposed a two-tier system. The lower tier would be simple applications, which would be delegated to officers. If a decision was not reached by a certain time permission would be given by default. The upper tier would be more complex applications, which would undergo the standard scrutiny described in Chapter 1.

The government reply in two circulars (DOE, 1975, 1976) was that: (1) it was not the system but the way in which it was operated that was at fault; (2) permission should be granted unless there is a sound and clear-cut reason for refusal. The onus was therefore put on LPAs to show that development was unacceptable rather than on the applicant to show that it was acceptable. Significantly this positive advice came from a Labour government, and was endorsed by the incoming Tories in DOE Circular 22/80 which stated that:

> Local Planning Authorities are asked therefore to pay greater regard to time and efficiency, to adopt a more positive attitude to applications, to facilitate development, and always grant planning permission having regard to all material considerations unless there are sound and clear cut reasons for refusal.

This quotation seems an appropriate point to end on, since it emphasises that development control is not a them-and-us process but one in which stakeholders should operate in an actor-network situation. By and large development control is an appropriate and effective reconciliation process for managing change within the context of Land Use Plans. Given this endorsement, it is now time to see what impact the process has had in Chapter 4.

Further reading
in addition to references already highlighted in text and illustrative material
Carmona, M., Carmona, S. and Gallent, N. (2003) *Delivering New Homes: Processes, Planners and Providers*, London, Routledge, provides an in-depth review of how the planning system operates in the housing sector and discusses several areas of tension: land disputes; delays; discretion; design; planning gain; and coordination.

Plan making
These papers will give further depth and detail to the readings already highlighted in the chapter:

Cloke, P. and Little, J. (1986) The implementation of rural policies: a survey of county planning authorities, *Town Planning Review*, 57, pp. 265–84.
Davoudi, S., Hull, A. and Healey, P. (1996) Environmental concerns and economic imperatives in strategic plan making, *Town Planning Review*, 67, pp. 421–35.
Tewdwr-Jones, M. (1997) Plans, policies and inter-governmental relations: assessing the role of National Planning Guidance in England and Wales, *Urban Studies*, 34, pp. 141–62.
White, A., Littlewood, S. and Whitney, D. (2000) A new space for sustainable development: regional governance, *Town Planning Review*, 71, pp. 395–413.

Web sites are these days a sometimes overused source and there is a danger of picking up polemical information. However, the Web sites of central government and local authority planning departments provide a plethora of useful information, although Land Use Plans by their very nature will often contain large and bulky files often in PDF format.

Development control
A detailed history of development control is provided by:

Booth, P. (2003) *Planning by Consent: The Origins and Nature of British Development Control*, London, Routledge. Traces development control back through several centuries.

A detailed study of development control procedures is provided by:

Thomas, K. (1997) *Development Control: Principles and Practice*, London, UCL Press.

The following papers add greater depth and detail to the references already cited:

Claydon, J. and Smith, B. (1997) Negotiating planning gains through the British development control system, *Urban Studies*, 34, pp. 2003–22.
Ennis, F. (1996) Planning obligations and developers: costs and benefits, *Town Planning Review*, 67, pp. 145–60.
Glasson, B. and Booth, B. (1992) Negotiation and delay in the development control process: case studies in Yorkshire and Humberside, *Town Planning Review*, 63, pp. 63–78.
Glasson, J., Therivel, R., Weston, J., Wilson, E. and Frost, R. (1997) EIA: learning from experience, *Journal of Environmental Planning and Management*, 40, pp. 451–64.
Jones, C., Wood, C. and Dipper, B. (1998) Environmental assessment in the UK planning process, *Town Planning Review*, 69, pp. 315–39.

Sellgren, J. (1990) Development control data for planning research: the use of aggregated development control records, *Environment and Planning B: Planning and Design*, 17, pp. 23–37.

Web sites can provide lots of useful information on development control, notably the Web site operated by the weekly *Planning* magazine and by the Planning Inspectorate. The most useful databases in these sites may be subject to a fee; check if your university has an access agreement. It is the government's aim to have all planning applications available electronically by 2010.

4 Evaluating Planning Outcomes and Impacts

There are two main aims in this chapter. First, to examine the outcomes of planning in terms of actual changes on the ground. For example, the amount of land transferred from rural to urban use. Second, to discuss the outcomes and impacts of planning via a number of case studies, notably in three different landscapes. In addition the impact on house prices is highlighted. The choice of topics is deliberately not inclusive of all outcomes and impacts. The choice has been determined by three factors. First, the Town and Country Planning Act's main focus has been on the process of managing change from one land use to another, notably rural to urban, or by finding 'brownfield' land within urban areas. Second, research has tended to concentrate on these processes of land use transfer and their wider impacts and any evaluation should be based as far as possible on the rigour and integrity provided by refereed research in respected journals. Third, this is a book about *land use* planning, in other words how land use has been affected by planning decisions.

Although the chapter is divided into outcomes and impacts the discerning reader will also note that three methodological approaches are used by the research evaluated in this chapter. First, there are those numerical studies that use published data in order to assess the changing rate of land transfer to urban use. These can, however, only infer the effect of planning on these rates. So, second, some studies of urban containment use case study evidence to assess the link between planning policies and land transfer. However, once again the causal link is inferred, so a third type of approach uses studies that have attempted explicitly to examine the link between planning policies and changes on the ground.

Learning objectives

1 To understand why planning has produced certain outcomes and to begin to explain how these outcomes can be interpreted in different ways and their wider impacts.

2 To understand why planning decisions have had a spatial outcome and then to understand why planning has had not only outcomes in its immediate area but also wider impacts. For example, an outcome is where 100 houses are built in one area rather than 200 in another area. The impact is felt in higher house prices and different travel patterns.

3 Taken together, outcomes and impacts provide a more exacting evaluation than the evaluation of processes in Chapter 3, but they are only a precursor to the overall evaluation in Chapter 5. So the third learning objective is to put planning in the context of the socio-economic system in which it operates.

MANAGING LAND USE CHANGE OUTCOMES

Land transferred to urban use

Most studies in this genre use the more emotive phrase 'land lost from agriculture'. There is little doubt that a core public perception of planning is as a device to prevent such land loss, and in a review of several public opinion surveys Hall *et al.* (2004) have found strong support for policies preventing urban growth in the countryside. Such perceptions are fuelled by emotive statements from pressure groups like CPRE that an area the size of Norwich is being built over every year. Given this perception, perhaps the most surprising fact is that the government has never seriously collected data on the planning system in spite of its cost and huge impact. It did make some attempt in DOE Circular 71/74, 'Statistics of Land Use Change', which asked LPAs as part of their new post-1974 functions to collect land use change data arising from their planning work, but LPAs never really bothered to do so. Instead Ordnance Survey data have been used and an OS/DOE series has been published since the mid-1980s. These data will be considered at the end of this section, after an examination of other data sources that were used in three contrasting studies by Coleman, Best and Sinclair.

The first study was carried out by Coleman (1976), who used a comparison of data derived from the first land use survey of the 1930s with data from the second land use survey of the 1960s to produce the data shown in Table 4.1. The data showed a dramatic rise in the built-up area of nearly 50% in 30 years and a loss of around half a million hectares of farmland. This was offset by some moorland reclamation, but most of the moorland loss was accounted for by woodland. Coleman's findings were used, by her and other workers alarmed at evidence of world food shortages, to argue that British planning should become much stricter on new development in the countryside. In addition, she argued that planning was failing to maintain a clear division between town and country, leading to the creation of an unattractive and wasteful 'rurban' fringe. However, critics (Gilg and Blacksell, 1981) pointed out that the data had been collected by volunteers, often schoolchildren (including the author) and could not be considered to be very reliable

Table 4.1 Land use changes in England and Wales, 1933–63 (ha)

Category of use	1933	1963	Change	% change
Settlement	1,092,111	1,629,142	+537,031	+49
Farmland	10,683,961	10,183,168	−500,792	−5
Woodland	856,763	1,212,525	+355,762	+42
Moorland, etc.	2,359,433	1,939,079	−420,354	−18

Source: Coleman (1976)

In contrast Best (1977), using data from the Ministry of Agriculture's annual farm census, argued that the rate of transfer had fallen significantly from around 25,000 ha a year in pre-war England to around 15,000 ha per year in post-war England, as shown in Table 4.2. Best also argued that these lower land losses would continue, since much land had been lost in the 1960s owing to the need to rehouse slum clearance areas at modern lower densities, as shown in Figure 4.1.

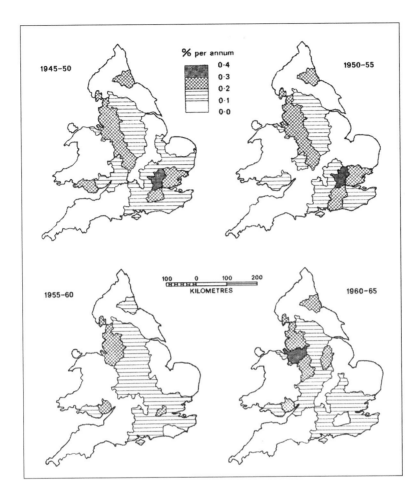

Figure 4.1 Regional conversions of farmland to urban use, England and Wales, 1945–65. *Source*: Best and Champion (1970)

Table 4.2 Annual average net losses of farmland to urban use, England and Wales (000 ha)

Period	Loss	Period	Loss
1922–26	9.1	1945–50	17.5
1926–31	21.1	1950–55	15.5
1931–36	25.1	1955–60	14.0
1936–39	25.1	1960–65	15.3
1939–45	5.3	1965–70	16.8
		1970–74	15.4

Source: Best (1977)

This shows that the North West in particular experienced rapid rates of land loss in the 1960s as 'Coronation Street'-type areas were replaced with tower blocks or houses in new towns. It also shows that there was a broad band of loss from London to Birmingham and on to Manchester and Sheffield/Leeds, leading to fears that a 'megalopolis' was forming along the line of the newly constructed M1 and M6 motorways. 'Megalopolis' was a term coined by Gottman, a leading geographer of the 1970s, to describe the growth of large cities linked together, notably Washington–New York–Boston in the United States. In contrast to the changes in the North West, the data also showed that the rate of land loss had begun to fall in the South East as development began to take place in market towns which offered plenty of potential for infill in their leafy suburbs. Best concluded that, in the future, standard densities of around 30 houses per hectare, infilling, and the use of derelict land, would aid a further reduction in the rate of land loss and that regional disparities would decrease.

Table 4.3 shows that this did in fact happen, with the rate of land loss falling from 8,700 ha in 1985 to 5,620 ha in 1995. In addition less rural land was being used, with the percentage of rural land used falling from 52% to 38%. Within urban areas, nearly 80% of development was on brownfield or redeveloped sites.

Table 4.3 Previous use of land changing to residential use, England, 1985–95

	1985		1990		1995	
Previous use	Ha	%	Ha	%	Ha	%
Rural	4,524	52	3,560	46	2,136	38
Urban	4,176	48	4,180	54	3,484	62
Total	8,700		7,740		5,620	

Source: Breheny (1998)

The debate was rejoined in the 1990s when the CPRE disputed data from the DOE derived from annual returns from the Ordnance Survey which purported to show that the rate of land loss had fallen to an amazing 5,000 ha a year by 1989. In contrast, Sinclair (1992), using a variety of data, produced a consensus figure, which showed a much greater loss, notably from the 1970s onwards as Sinclair's consensus figures and the DOE's figures increasingly diverged. However, both sets of figures showed a decline in the rate of land loss. Thus there is general agreement that the rate of land loss in the 1960s was between 16,000 ha and 18,000 ha. In the 1970s, however, the DOE figure fell to 10,000 ha by the end of the decade, but Sinclair's figures after rising to around 25,000 ha in the mid-1970s fell to 18,000 by 1980. In the 1980s the DOE figures fell to below 5,000 ha but Sinclair's figures only fell to around 12,000 ha. Over the whole period, Sinclair argued, his figures showed a total land loss of 705,000 ha compared with the official figures of 525,000 ha.

Sinclair explained the differences largely because of unreported changes by farmers in the agricultural census, and went on to argue that DOE policies regarding a more lenient approach to land release were flawed, since they were based on complacent and unreliable data. He concluded that if current trends continued, the urban area would increase from 15% to 25% by 2050 or the equivalent of a town like Norwich every year.

However, official data continued to show much lower rates of land loss than Sinclair, partly because by 1992 47% of new housing was being built on previously developed land, compared with 38% in 1985 (Cm 3016, 1995). Between 1992 and 1996 the total land changing to residential use was estimated to be 6,325 ha and then from 1997 to 2001 to be 5,470 ha (Land Use Change Statistics 1997–2001, www.planning.dtlr.gov.uk/lucs17/index.htm). Since around 36% of this change was from agricultural land the estimated land lost from rural areas was probably only around 2,000 ha.

However, this period also witnessed one of the lowest levels of house building since the 1930s, as shown in Figure 4.2, largely caused by the virtual elimination of council houses in 1979 when Mrs Thatcher came to power. Given that Figure 4.2 shows a drop from over 400,000 new houses in the late 1960s to only just over 150,000 houses in 2001, it is not surprising that the rate of land loss has dropped. Another factor contributing to the reduction was the high density of new housing developments, which averaged 25 dwellings per hectare between 1997 and 2001 and had remained much the same since 1991. There were, however, big differences in the rate of brownfield development, varying from 90% in London, 61% in the North West and 55% in the West Midlands to lows of 37% and 38% in the South West and East Midlands. By 2003 the percentage of development on brownfield land had risen from 59% in 1999 to 66%, with densities rising from 25 dwellings per hectare to 30 per hectare (Office of the Deputy Prime Minister, 2004)

In a major contribution to the debate Bibby and Shepherd (2001) used the DTLR's Land Use Change Statistics derived from Ordnance Survey data and combined it with digital boundary data showing the configuration of urban land at 1991. This allowed them to analyse the type of land on which housing had been built between 1985 and 1992 by three categories. The first category, the proportion of housing built by urban expansion, in other words at the urban fringe, was found to be common only in an arc stretching from Bristol to Cambridge and is probably a reflection of market town expansion unfettered by Green Belts. The second category, the percentage of housing built on infill sites, was most prevalent in the big cities, where the girdle of Green Belts constrains new housing within its squeeze. Finally, the third

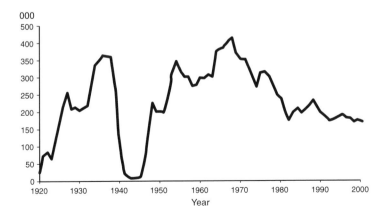

Figure 4.2 Total houses built in Britain (DETR statistics), 1920–2001. *Source*: adapted from *The Times*, 3 May 2002, p. 8

category, the proportion of housing built on rural sites, was dominated by the South West, East Anglia and the North.

It would be a mistake, however, to think that all urban infilling was on brownfield land, for as Bibby and Shepherd (2001) have reported, nearly a third of the land used for dwellings in urban areas between 1985 and 1992 was on greenfield sites. Conversely, around a fifth of development in rural areas was on brownfield sites. Bibby and Shepherd also note that the percentage of brownfield land used to be a lot lower, at 43%, in the late 1980s than the proportion reached by the late 1990s of nearly 60%. This percentage should be achievable for several years to come, since in September 2002 the National Land Use Database (*Planning*, 20 September 2002) reported that there was enough brownfield land for 920,000 houses distributed as follows: North East (48,000); North West (138,000); Yorkshire and Humberside (95,000); East Midlands (79,000); West Midlands (71,000); East Anglia (109,000); London (149,000); South East (119,000); and South West (107,000).

Also for the future Adams (2004) studied the reaction of the small number of large-volume house builders who now dominate the house-building scene to the brownfield target. Some had already appreciated the far-reaching implications of the regulatory changes of recent years but the majority still saw brownfield land as a specialised part of their business. There is thus a clash between structure and agency, with most commentators seeing the issue as how house builders will react to the new regulatory structure. Ultimately, the success of the 60% target will depend on the willingness of house builders not merely to accept it with reluctance but rather to devise innovative ways to meet the policy objectives with some style. Thus policy makers are not distant spectators but have a vested interest in the rapid development and dissemination of brownfield competences, even if that can be achieved only by significant restructuring of the speculative house-building sector

The next section examines the related issue of urban containment, not from the perspective of disputed databases but from various case studies that have attempted to examine where urban environments have been built and whether population has followed new housing or vice versa.

Urban containment

Urban containment was one of the key aims of the 1947 Act and remains so today, no longer to conserve farmland but to preserve the countryside and to provide more sustainable urban forms based on higher-density living with less need for travel. These aims, however, produce deep conflicts in a society based on mobility, as all the case studies will show.

Peter Hall's classic study

The first and still the classic study of containment remains that carried out by Peter Hall *et al.* (1973) in the early 1970s and published in a massive tome, *The Containment of Urban England*. Their main thesis remains valid today, namely that decentralising forces in the population and the workplace are being opposed by containment policies and that this was a long-term reversal of the centralisation of the industrial revolution in the nineteenth century, as shown in Table 4.4. This demonstrates the population being more mobile, decentralised first, followed by employment. In other words, people moved to the edge of cities or beyond, but employment stayed in city centres before decentralising later on. This decentralisation poses a great threat to established interests in rural areas and is really the key planning problem in England. Namely, the desire to preserve the countryside and maintain capital built up in the cities against the need and desire of people and employers to locate in modern low-density environments close to modern transport links.

Table 4.4 Centralisation and decentralisation in the twentieth century

Stage/approximate dates	Population	Employment
1 Pre-1900	Centralising	Centralising
2 1900-1950	Decentralising	Centralising
3 1951 or 1961 onwards in larger areas	Decentralising absolutely	Decentralising relatively
4 1961 onwards, in London and Manchester only	Loss in both core and ring; gain in peripheral areas	Loss in both core and ring; gain in peripheral areas

In addition to this core finding the Hall team (1974) came to three big conclusions. First, urban areas had been contained and the threat of a megalopolis from London to Manchester, although a potential threat, was a long way from reality in 1960, as shown in Figure 4.3, although rapid population growth was beginning to infill the gaps between the highly built-up areas. It was found that private owners had decamped over Green Belts into villages or expanded smaller towns, with council tenants being rehoused in tower blocks. It was argued that this was a civilised British form of apartheid, with Tory shires being protected by planning from working-class incursions and Labour voters.

Second, there had been a growing separation of work from residence. Third, there had been a severe inflation of land and property values. For example, the 'land cost' of a new house rose from just over 10% of the price in 1960 to around 25% by 1970. The net result was that planning was creating two types of environment. First, middle-class housing estates built down to what people could afford in places remote from existing services. Second, working-class housing in tower blocks on the edge of towns. As Peter Hall (1974) put it so graphically over 30 years ago:

It certainly was not the intentions of the founders of planning that people should live cramped lives in houses destined for premature slumdom far from urban services or jobs, or that city dwellers should live in blank cliffs of flats far from the ground without access to play space for their children. Somewhere along the way a great deal was lost, a system distorted and the great mass of people betrayed.

(p. 407)

Or as Evans (1992) put it so eloquently 20 years later: 'Rabbit hutches on postage stamps.'

In a partial update from secondary data Hall (1988) pointed out four great changes, which were causing profound changes in the geography of contemporary Britain

1　Zero population growth but a marked redistribution of population into the so-called Golden Belt (the arc from Cornwall to Norfolk) and the Golden Horn (a spanner-shaped region from West Sussex around the coasts of Kent and East Anglia to Lincolnshire). In these two regions there is and will continue to be a voracious demand for conversion of rural land to urban purposes.

2　A shift of land out of agriculture as shown in Table 4.5 which demonstrates that around 33% of land would be surplus to agriculture if farm productivity rose by 2.5% per year and food

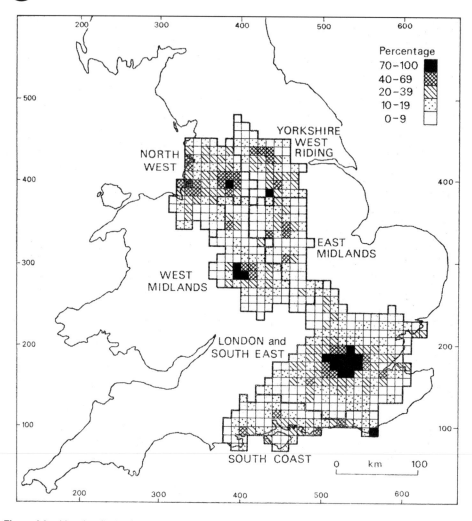

Figure 4.3 Megalopolis: land use, showing degree of urbanisation *c.* 1960. *Source*: Hall (1974)

demand only rose to 103 on an index of 100 in 1985. By the time 2000 came around 10% of farmland was 'set aside' under the Common Agricultural Policy and the catastrophic fall in farm incomes indicated that farmers were producing more than the market could bear. Indeed, around 20% of the cereal harvest is exported. Thus since the 1980s there has been a paradox of substantial areas of wasteland near the hearts of the cities, substantial areas of potentially surplus farmland, while demand for land is buoyant at the urban fringe where the political lobbying of the NIMBY crowd is at its most intense.

3　A shift away from manufacturing to high-technology services. For example, between 1963 and 1978 employment in manufacturing fell from 35% to 25% of the work force while employment in retail and catering rose from 17% to 20%, in financial services from 4% to 9% and in community services from 21% to 29%. These employment sectors are more mobile and more likely to be attracted to high-quality semi-rural environments.

4   A demand for out-of-town warehousing and retailing, which saw the area of such development rise from 28 million ft² to 108 million ft² between 1980 and 1987.

Table 4.5 Percentages of land forecast in 1985 to be surplus to agriculture by 2000

	Annual rate of productivity growth/Level of self-sufficiency					
	1.5		2.0		2.5	
Demand 1985	75	90	75	90	75	90
100.0						
103.0	22.7	7.2	28.4	14.1	33.4	20.1
104.0	22.0	6.4	27.8	13.3	32.7	19.3
105.0	21.2	5.5	27.0	12.5	32.1	18.6

Source: Hall (1988)

Hall argued that the 1980s witnessed a fundamental shift in the way in which we use our environment. In particular a massive move away from industrial jobs in the inner cities towards service employment, in out-of-town locations close to motorways, airports and new railway stations based on high-speed trains. However, he also argued that the planning system had not reacted to these changes. Instead it was still locked into an ethos of 'urban containment' and maintaining a settlement pattern based on the industrial revolution when what was needed was a new vision of a low-density road-based environment based on using up surplus farmland. Pennington (2000) has argued that this did not take place because a coalition of pressure groups and public-sector bureaucrats benefited from keeping containment at the core of the planning system. However, many extra-urban developments have been allowed, and to some extent the world has turned upside down, with new developments taking place in spacious countryside locations with generous tree planting and former central areas being transformed into leisure environments like marinas.

This transformation has to a high degree been related to the motorisation of society, with the great majority of journeys being completed by motor car or lorries. This process has taken place in all advanced countries, and planning policies have had to accept it as part of modern life. A number of anti-road pressure groups have argued that planning policies should have been more restrictive and maintained higher-density cities in which the road system would have become clogged up. Compact cities have indeed become the aim of many planning policies, although as Burton (2002) has pointed out there is little evidence to support the claims made in their favour, partly because there are no agreed methods to measure compactness or how it would reduce traffic. Moreover, work by Breheny (1995) and Stead (2001) demonstrates that there is no strong relationship between land use patterns and transport patterns. Stead has calculated that socio-economic characteristics explain two-thirds of travel behaviour, with land use patterns accounting for the other third.

There is thus a three-way relationship between travel patterns, land use patterns and socio-economic characteristics. The challenge for planners is to produce a land use pattern that provides mobility without sacrificing too much land and creating too much noise and pollution. They have attempted to do this by zoning land uses and the results have been patchy. In some areas transport is heavily congested and is a major pollutant but in other areas people enjoy the comfort and flexibility of road travel.

Green Belts

Green Belts play a key role in containment policy by providing a broad zone where development is strongly discouraged. However, Green Belts did not form part of the armoury of the 1947 Act, even though the idea began in the 1930s, notably with Abercrombie's plans for a Green Belt around London. Even today Green Belts have no specific legislation, and derive their power only from government advice and being contained in Development Plans.

The first advice came with a circular (42/55) from the Ministry then in charge of planning, the Ministry of Housing and Local Government, which set out three roles:

1   To restrict the sprawl of large built-up areas.
2   To prevent the merging of adjacent towns (e.g. Cheltenham and Gloucester).
3   To preserve the setting of a town (e.g. Oxford).

In spite of an attack by the Tory right wing in the early 1980s, Green Belts emerged stronger and new aims were added. For example, DOE Circular 14/84 added two more roles:

4   To safeguard the surrounding countryside from further encroachment.
5   To assist urban regeneration.

These five aims were endorsed in PPG2 in 1988 and in 1995 a new sixth aim was added when PPG2 was revised:

6   To provide access to the open countryside for the urban population.

The area protected by Green Belts experienced a massive rise in the early 1980s as a result of the first generation of Structure Plans, with the total area in England growing from 700,000 ha to 1,800,000 ha. By the turn of the century most major towns and cities had a Green Belt.

The main power of Green Belts comes from the policy guidance they give to development control through approved Development Plans. However, this does not mean development is forbidden, only that a special case has to be made. Munton (1983) and Elson et al. (1993) found that special cases have in particular been made for institutional uses, notably hospitals, universities and schools, and for leisure uses such as golf courses, country parks and 'horsiculture' (pony stables and riding schools). Housing has also been allowed, notably at the boundary of the Green Belt where the boundary's accuracy has been queried. The resulting landscapes tend to be attractive, but perhaps over-manicured. In contrast, farmland tends to be run down in the face of urban-fringe vandalism, fly tipping and general neglect in advance of a more lucrative land use.

The presence of the Green Belt increases its desirability as a residential environment and so not surprisingly developers are continually testing the resolve of planners and the inspectorate to uphold the policies. One such test came in the late 1980s when a number of developers ganged together to form Consortium Developments. They applied for permission to build a new town at Tillingham Hall in the Essex Green Belt. The application created an outcry from conservation groups, notably the CPRE, and the application was rejected both by the LPA and by the Minister on appeal. Further applications were then made in Kent and in Hampshire. Most of these were rejected on appeal, but one ironically named Foxley Wood was given a potential green light. However, at this stage, the early 1990s, Mrs Thatcher came to see the votes to be kept in the Green Belt and confirmed that Green Belts would continue to be protected from major residential development.

Given the pressure for development in Green Belts it is not surprising that the pressure needs to be released somewhere else, notably just beyond them as they are leapfrogged by developments. Herrington (1984) juxtaposed population change between 1971 and 1981 and Green Belts to demonstrate that most population change had indeed taken place either at the margin or just beyond the Green Belts. The exception was the North, where much change had taken place in the Green Belts, though this may reflect permissions before the belts were extended and also the lack of alternative land in the North West.

Keyes (1986) has shown how a change of policy had an effect in the Essex Green Belt. Between 1980 and 1984 83% of applications were refused inside the Green Belt, but when the policy was relaxed in 1982 the rate of refusal fell from 96% in 1980–82 to only 55% in 1982–84. Outside the Green Belt the refusal rate remained exactly the same at 25%. Most applications were made at the border of the Green Belt with the built-up area, rather than well inside it, suggesting that developers were testing whether the boundary could be defended as being drawn in the right place. However, Keyes's work took place before Green Belt policy was strengthened again in 1984. Indeed, by the late 1990s only 3% of all new dwellings were built in Green Belts and 60% of these were on brownfield land. Residential development in the four years from 1997 to 2001 accounted for only 0.02% of Green Belt land (Land Use Change Statistics 1997–2001, www.planning.dtlr.gov.uk/lucs17/index.htm).

Green Belts have thus been very successful in their main aim of restricting residential development but access is restricted to landowners or private members' clubs, e.g. golf courses, and they have only been partly instrumental in diverting growth to the inner cities. Although they are popular with the voters, two groups of planners, the RTPI and the TCPA, have queried whether Green Belts should be reduced and the restrictive policy relaxed in the face of rising house prices and severe land shortages for development, notably in south-east England. Thus it is now time to proceed to a case study of south-east England to explore more fully the tensions between landscape protection and residential development.

The south-east of England as a case study of the land use change debate

The nub of the planning problem in the region is that it is the economic driver of Britain, but also one of the most crowded and congested parts of Britain, with also some of the best landscapes and farmland. It is also the heartland of opinion making by the so-called 'chattering classes'. Indeed, in the 1980s the *Sunday Times* newspaper ran a 'Memo to Maggie' column in which it repeatedly called on Mrs Thatcher (Maggie) to relax planning controls in the South East. However, by the mid-1990s, the *Times* newspaper was running a campaign to save our countryside from urban growth. This reflected a massive split in the Tory Party between its business side and its grass roots of house owners. There was also a political dilemma for New Labour from its election in 1997, which had depended on Tory voters deserting to the

Labour or Liberal Democrat parties. All parties, however, had to promise strong controls on development while at the same time offering economic growth.

The problems in this balancing act were brought to a head by the dramatic expansion of Vodafone on the back of the mobile phone revolution of the late 1990s. Vodafone grew from a staff of 100 in 1982 to over 3,500 staff in 1999. These were, however, accommodated in 57 buildings in its home town of Newbury. The company applied for planning permission to construct one building in order to reduce travel time between its various locations. This was somewhat ironic, given the revolution in communications brought about by its main product, mobile phones. The new building would be built on a greenfield site covering 12 ha and would release over 50,000 m^2 of office space and potentially 4,000 new jobs, with a consequent demand for new housing. The issue was also complicated by a long-standing conflict about the Newbury bypass, which had recently reduced traffic in the town but at the expense of important wildlife sites and landscapes. Newbury thus illustrates how the dynamics of a new industry can fundamentally change a small market town in little more than a decade and how planning was forced to allow these changes to take place by permission being granted in 1999. It also highlights the dilemma faced by many market towns. Finally, it not only illustrates the need for regional planning, but also highlights how regional plans may not foresee radical changes in technology and wealth creation. For example, mobile phones were transformed from an expensive business device to a ubiquitous item owned by virtually everybody in just a few years in the late 1990s.

Nonetheless, regional plans provide a forum for debating the issues. In the South East the regional planning process focused around different housing forecasts in the region, as shown in Table 4.6. At the outset, SERPLAN, the regional planning conference of all the south-eastern counties outside London, produced a forecast of 666,00 houses needed between 1996 and 2016, or 5,700 houses lower a year than the then building rate of 39,000 houses a year. This reflected strong anti-growth sentiment among planners, councillors and voters. However, in 1998 the panel report after the public inquiry chaired by Stephen Crow, the former chief planning inspector, caused an outcry when he dramatically raised the forecasts to over a million houses. Only the Town and Country Planning Association and the house-building and development lobbies welcomed the proposal. Two years later, in a politically astute move, the Secretary of State took a middle figure of 860,000 or 4,000 houses a year more than the current build. In order to appease the NIMBY lobby, however, he raised building densities to between 30 and 50 dwellings per hectare compared with the previous average of 24 per hectare and set a target of 60% of all new dwellings to be on brownfield land.

Table 4.6 also shows strong sub-regional differences, with the most rural counties taking the biggest forecasts, notably Hampshire, Buckinghamshire and Essex, with major expansions at Milton Keynes and in Kent. Further pressure was put on the South East with another announcement in July 2002 that the South East needed four new runways at Heathrow and Stansted and probably a new airport on the north Kent coast to cope with the predicted doubling of demand for air travel in the next 30 years.

In spite of these pressures the *Sunday Times* was able to report on 30 June 2002 that only 82% of the houses expected to be built by certain counties in the South East RPG were actually built in 2001. Most areas were far below the target, as in West Sussex, where there had been a campaign to argue that the county is 'full up'. Only Surrey was on or around the target and perversely the Isle of Wight was well over target. The reaction to these shortfalls, newspaper editorials (see Box 4.1) and dramatically rising house prices as construction dropped to the lowest level since the end of the Second World War, was to relax planning controls in July 2002, as shown in Box 4.2. However, the government also announced that Green Belts would still be protected in recognition of their great popularity.

Table 4.6 Disputed housing forecasts in the south-east of England

County	SERPLAN	SOS	County	SERPLAN	SOS
Bedfordshire	43,100	55,700	East Sussex	37,500	48,400
Berkshire	53,300	68,800	Hampshire	101,900	131,600
Bucks	64,300	83,000	Hertfordshire	50,900	65,700
Isle of Wight	9,500	12,300	Kent	99,700	128,800
Oxfordshire	41,600	53,700	Surrey	34,900	45,000
West Sussex	45,400	58,700	Essex	83,900	108,300
Total	666,000	860,000			

Original SERPLAN proposal	666,000	(33,300 houses per year)
After public inquiry (Crow) proposal	1,100,000	(55,000 houses per year)
Secretary of State proposal	860,000	(43,000 houses per year)
Current build	780,000	(39,000 houses per year)

Source: The Times, 8 March 2000, p. 4

Box 4.1 An editorial in the *Sunday Times*, 30 June 2002

Gordon Brown's support for a housing boost and world-quality transport system in south-east England is a welcome change from the days when governments tried to run Britain's economy at the pace of its slowest regions. Holding back the most dynamic part of the country to pay political lip service to Scotland, Wales and northern England never made much sense, cost billions of pounds and has not worked. While both parties hobbled Britain's most vibrant wealth-creating area, the UK's continental rivals proved that high economic expansion can go hand in hand with a good environment and quality of life.

A new study puts Frankfurt, Munich, Amsterdam, and Paris well ahead of London in their overall quality of life and they are at the heart of Europe's most prosperous areas. Faster economic growth, which depends on a bigger, more mobile population, could bring added environmental benefits for London and the South East. But that cannot happen if planning consent takes eight years to come, as it did for London's fifth terminal at Heathrow airport, and special interest groups fight every development on principle as far as the House of Lords. Nobody wants to turn the South East into a megalopolis. On offer, as the Chancellor recognises, is the prospect of making it an even stronger hub of Britain's prosperity without destroying its green belts and covering the Home Counties with housing developments. Quality of life matters but families who cannot afford to get on the first step of the property ladder have rights, as well as those who live in detached houses and suburban semis within reach of rolling fields and pretty villages.

High-density building is needed to accommodate a larger population on 'brownfield' sites. Other countries build good high-rise flats; Britain can too. House building nationally is at a seventy-year low. The South East's urban regeneration would transform the picture, along with better roads, railways and air services. Only then can the two-thirds who live in the rest of Britain share in the greater wealth created by the third who live in the bottom right-hand corner.

Box 4.2 Policy changes announced in July 2002

1 200,000 new homes to be built in four development zones to the north and east of London at Ashford, Thames Gateway, Bishop's Stortford and Milton Keynes.
2 All councils in London and south-east England to make up a shortfall of 20,000 homes annually.
3 Cheaper homes for key workers, e.g. teachers and nurses.
4 New communities in Milton Keynes, Greenwich and Hastings and outside the South East near Birmingham (East Ketley), east Manchester, King's Lynn and Leeds (Allerton Bywater).
5 Planning powers to be taken from county councils and given to unelected Regional Development Agencies.
6 New targets for local authorities to hasten planning approval.
7 Local authorities to set up business development zones with virtually no planning controls.

Area	Current build	New houses needed
South East	24,000	43,000
London	13,000	23,000

Source: The Times, 19 July 2002, p. 4. For an update see the current Web site dealing with the 'Sustainable Communities' programme.

A year later, in July 2003, the government announced its 'Sustainable Communities' plan to build 200,000 new homes in four sub-regions based around Ashford, the Thames Gateway (east London), Stansted–Cambridge and Milton Keynes and the south Midlands. Public money was made available to help kick-start the process. It was also suggested that if local planning authorities attempted to veto the process special organisations would be set up to assemble land along the lines of the new towns in the 1950s, which had powers to buy land and give themselves planning permission.

The overall problem is that voters in leafy green constituencies have obtained their 'positional good' or place in the sun. Strict planning controls not only protect their high-quality environment (NIMBY) but also, in line with Mark Twain's famous aphorism that 'land is the one thing that isn't made any more', makes their property more valuable because no more properties are being built. If one political party relaxes the controls, the other parties will pounce on their disaffection. This cannot be allowed, since the South East with its electoral mix is where elections are won or lost. However, a survey by Cambridge Architectural Research (2004) of 1,400 residents in the region found that though 46% did not want many more houses built, 28% thought that more houses should be built.

In more depth a fascinating study by While et al. (2004) showed how a pro-growth agenda may come about even in an area like Cambridge with a strong and highly articulate middle class. Cambridge with its world-class university is clearly a highly desirable location for the new service industries but the city has had a strong Green Belt policy and is situated in an area of good-quality agriculture. In spite of this, three strategies emerged in order to provide a pro-growth agenda. First, the local mode of regulative planning was transformed. From the need to preserve the traditional setting of the last true 'university town' to one where the need to accommodate the high-tech phenomenon of the 1990s in the most suitable area was an issue of national importance to Britain's economic future. Second, territorial politics and a new space for governance emerged as the county council became more important as an actor than the city council with the revival of regional planning in the 1990s. Third, appeals were made to the central state that Cambridge, as a prosperous area, did not benefit from UK or European funding. Ironically the potential for growth was so great that it should be allowed to grow in order not only to keep pace with its competitors but also to provide the funds for aid to less prosperous regions.

Other key features of the process were that from the outset it was made clear that major development would be located outside the historic city. Pro-growth lobbyists also argued that planned growth was better than development by stealth or opportunistic planning applications in haphazard locations. Next, the county council became a key broker in the pro-growth campaign, which undermines the view that all shire counties are anti-growth, whereas in this case the city council was more anti-growth. Finally, the authors conclude that it may be unwise to suggest that the state, especially the regional and local state, is becoming marginal or 'hollowed out'. Instead they are drivers in their own right rather than merely enablers.

In a review of these issues Murdoch (2000) employed the work of Foucault and in particular his notion of governments conducting debates like an orchestra in an attempt to shape behaviour according to particular sets of norms for a variety of ends. In this process governments use networks to conduct the debate but these networks are precarious and susceptible to disruption by external forces and there are also tensions between time and space. Murdoch found that the process of planning in the South East was messy and heterogeneous with a mix of technical assessments, for example household projections, and political judgements. He concluded that short-termism won, with Prescott (the Minister in charge of planning) stressing the annual nature of the forecasts, that space won out over time by moving power to Regional Planning Guidance and that environmental issues allied to NIMBY politics reduced forecast housing projections.

Great Britain: facts, figures and two conceptualisations

The situation in the South East is complicated by the need to siphon growth if possible from the South East to the rest of Great Britain and to balance growth in the rural areas with decline in the inner cities. The problem of uneven regional and urban–rural growth can be shown by figures from the population censuses of 1971, 1981 and 1991. These show that rural areas experienced a population increase of 1.8 million between 1971 and 1991, and a percentage rise of 16.9% compared with 3.9% for England as a whole.

Champion (1993) has examined these figures in more detail, as shown in Table 4.7. This shows that the southern half of Great Britain experienced the greatest growth both absolutely and relatively, notably East Anglia and the South West. In contrast, the North and Scotland experienced population loss. A more detailed breakdown by area shows that the biggest percentage increases stretched in an arc from Cornwall to Norfolk. This replicates the population geography of England before the industrial revolution, and suggests a return to a pattern based on the best climates and landscapes of Britain. Unfortunately, most transport investment has been based on lines radiating from London, notably north–south links, and there are few good communications along this new dynamic arc.

The underlying components of these changes are interesting. Natural change although accounting for most of the total change is, however, strongly concentrated in the South East and the West Midlands. In-migration dominates in East Anglia and the South West, where

Table 4.7 Components of population change, 1981–91, by region and change, 1981–2001

	1981–91				Change 1981–2001	
	Natural change		Migration, etc.			
Region	000	%	000	%	000	%
East Anglia	29.2	1.54	167.3	8.83	534	11
South West	−6.1	−0.14	348.1	7.94	545	12
East Midlands	76.4	1.98	96.7	2.51	319	8
South East	451.6	2.65	95.5	0.56	758	10
Wales	25.0	0.89	48.0	1.71	90	3
West Midlands	138.3	2.67	−70.1	−1.35	81	2
Yorks+Humberside	65.1	1.32	−29.4	−0.60	46	1
Northern	18.1	0.58	−51.3	−1.64	−121	−5
North West	87.1	1.36	−169.4	−2.62	−211	−3
Scotland	27.0	0.52	−107.2	−2.07	−118	−2
Great Britain	912.2	1.66	328.1	0.60	2,432[1]	4

[1]UK

Source: Champion (1993); *Planning*, 11 October 2002

it is particularly strong, but in the North and Scotland out-migration has in some cases caused net population losses. A more detailed breakdown shows that the most dynamic area with strong birth rates and in-migration lies in a spine through the centre of southern England and the Midlands. Around this spine another dynamic area relies on in-migration. The surrounding rural areas rely entirely on in-migration. In contrast, the big cities rely on natural change to offset out-migration.

Champion also analysed the changes by the type of district. This demonstrated that the biggest absolute gainers were the areas of urban and mixed urban/rural environments typified by districts to the south-west of London. In these areas traditional market towns offer an attractive environment for both employers and employees. In contrast, the principal cities, outside London, offer poor environments and they experienced both the biggest absolute and the biggest relative losses. Table 4.8 show the overwhelming popularity of areas with a good rural or semi-rural environment and this poses a tremendous challenge for planners. Namely, how to manage growth in these areas without spoiling them, so that the process of inner-city decline does not repeat itself until there is nowhere left in Britain to 'mess up'. This problem is made all the more ironic since a number of 'quality of life' surveys (Champion, 1989) give higher scores to the North, largely because of the less congested and lower house price environment. When the 'North' becomes the 'new black' a whole new set of planning problems may emerge. Meanwhile the long-term problem of the overheated South East remains a core planning problem.

Employment figures were also up in rural areas by 13% between 1981 and 1991, compared with a rise of only 1.2% for England as a whole. This replicated to a large extent changes in the 1970s when the big cities lost around 10% of their work force while 14 rural areas or new town areas experienced a more than 20% rise in jobs. For example, Bracknell was up by 28% and Thetford up by an amazing 30% (Champion, 1993)

The 2001 census updated population change from 1991 to 2001 and though at the time

Table 4.8   Population change by top and bottom areas, 1981–91 (%)

Top		Bottom	
Counties			
Cambridgeshire	13.7 (PR)	Shetlands	−14.6 (RD)
Buckinghamshire	12.3 (PR)	Western Isles	−6.7 (RD)
Cornwall	11.4 (RE)	Merseyside	−5.3 (ID)
Dorset	10.7 (RE)	Strathclyde	−4.9 (ID)
Northamptonshire	10.3 (NT)	Tyne and Wear	−2.6 (ID)
Districts			
Milton Keynes	42.0 (NT)	Shetlands	−14.6 (RD)
Kincardine/Deeside	26.7 (NS)	Clydebank	−12.2 (ID)
Forest Heath	24.1 (PR)	Clydebank	−12.2 (ID)
Wokingham	21.3 (PR)	Knowsley	−11.2 (ID)
Gordon	21.0 (NS)	Inverclyde	−9.5 (ID)
Huntingdon	19.8 (PR)	Liverpool	−8.2 (ID)
Redditch	16.6 (NT)	Salford	−7.9 (ID)
Bracknell	16.4 (PR)	Dundee	−6.8 (ID)
South Hams	16.2 (RE)	Western Isles	−6.7 (RD)
Northampton	16.1 (ET)	Monklands	−6.7 (ID)
Selby	15.9 (CO)	Manchester	−6.5 (ID)
Peterborough	15.7 (ET)	Port Talbot	−6.4 (ID)

Probable explanations for change attributed by author: *PR* Pleasant rural, *RE* Retirement, *NT* New town, *NS* North Sea oil, *ET* Expanded Town, *RD* Rural decline, *ID* Industrial decline, *CO* Coal mining.

Source: Champion (1993)

of writing no detailed analyses like Champion's work had been replicated, some points can be made. For example, Figure 4.4 demonstrates that people continued to leave the big cities, except for London, which registered growth, probably because of the large number of foreign workers entering the country. Strangely, the London Green Belt registers several areas of decline, probably reflecting the severity of restraint policies in these areas. There is still a major north–south divide, with only a few rural areas registering increases. The dominant pattern is, however, a belt of growth stretching in an arc from East Anglia through Milton Keynes to Bristol and then down to south Devon and Cornwall. Much of this growth has been in and around towns but some rural areas have also registered large increases, albeit in percentage rather than absolute terms. Clearly planning permission is being given in rural areas but we can only speculate about the detail. The main point to make from this is that it is scandalous that we know so little about the detailed pattern of land use changes mediated by the planning system, given the costs involved and the ferocity of the debates surrounding new building in the countryside.

In summary, Moss (1978) has provided a conceptualisation of the process of population relocation, as shown in Figure 4.5, which still remains valid. In particular, it emphasises the massive decentralisation of cities on the basis of a network of expanded towns and settlements threaded along the motorways. In contrast many villages are static or in decline as second-home or retirement dwellers move in, whilst the inner cities experienced severe decline at least till their partial renaissance from the late 1980s.

A more complex conceptualisation of how planning policies now see the countryside and

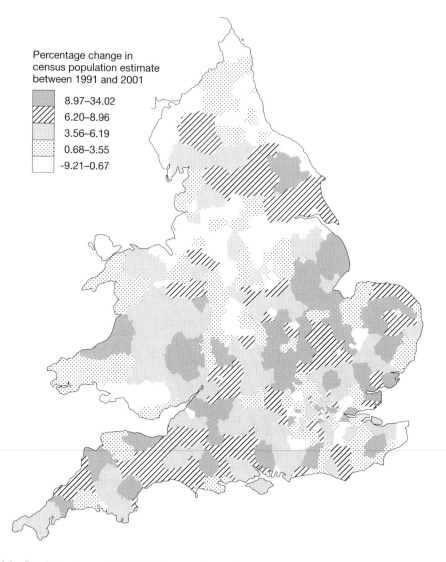

Percentage change in
census population estimate
between 1991 and 2001

8.97–34.02
6.20–8.96
3.56–6.19
0.68–3.55
-9.21–0.67

Figure 4.4 Population change, 1991–2001. *Source*: adapted from Office of National Statistics map. *Source*: www.statistics.gov.uk/census2001

nature has been provided by Murdoch and Lowe (2003), who have argued that countryside politics in England must now accept a revised view of nature. No longer can nature be seen as simply a static entity, one that needs to be protected against destructive human actions; rather, it should be conceptualised as fluid and dynamic, and as closely bound into economic and social processes of change. The focus of environmental politics should therefore shift from the *preservation* of 'pure' nature to the *sustainability of* 'heterogeneous' relationships, wherein social entities form 'partnerships' of various kinds with natural entities. The goal of environmental policy thus becomes the promotion of these partnerships so as to ensure that

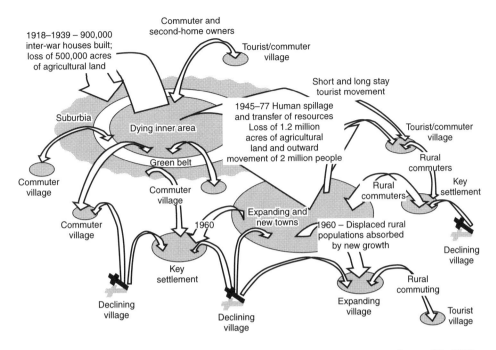

Figure 4.5 Conceptualisation of population and settlement change around major cities. *Source*: Gilg (1996 from Moss, 1978)

natural entities become sympathetically incorporated into the networks that increasingly connect urban and rural spaces.

This new viewpoint is derived from a study of the history of the CPRE from its inception in the 1920s till today. The CPRE began with modernist advocacy of planning as orderly progress and its support for the Town *versus* Country Planning Act of 1947. Two networks emerged. A national elite linked with national figures and local activists in pressured rural areas resisting growth. Paradoxically, the protection of pressured rural areas made them even more attractive and thus more people came to live there so that areas of counter-urbanisation then attracted more members to the CPRE.

In the 1980s the hegemony of the CPRE as *the* rural pressure group was threatened by the rise of Friends of the Earth and other 'green' groups. The CPRE moved towards a more holistic stance and replaced Preservation with Protection in its title. This happened only at the national level, though, where trade-offs were acceptable in the context of sustainable development. The 4.4 million extra houses debate of the 1990s gave the CPRE a major chance to influence policy. It employed Bramley to dispute the circularity of the 'predict and provide' approach. He found that local projections did have the defect of a self-fulfilling prophecy but not national figures. The CPRE moved the attack to environmental capacity and stresses on water, minerals and waste. Alongside this it argued for brownfield sites and the sequential approach. This was endorsed by DETR reports, the Urban Task Force and finally by PPG3 in 2000. Thus there is the paradox that the CPRE has traversed the urban–rural divide by becoming involved in the urban renaissance debate.

Thus the CPRE has become ecologically involved and the urban–rural divide is increasingly being replaced by an intricate urban–rural network. If we accept these arguments then at long last the divide between town *and* country planning set up in the 1947 Act should now be weakening rapidly in that the two areas are being examined together in terms of national and regional housing forecasts. However, the sequential approach in which inner-city sites, brownfield sites and any other urban site must be examined for development before greenfield applications can be countenanced demonstrates that the essential two pillars of the 1947 Act still hold sway. As much building as possible should take place in urban areas and the countryside should be protected. The second half of the chapter highlights how these twin pillars have been very resistant to change, but also highlights a number of attempts to undermine these formidable edifices.

Summary

The transfer of land use from one category to another is the primary function of the land use planning system. The most important transfer is from farmland to urban land and this has been especially important in a small crowded island like Britain, notably in one of its most crowded regions, the South East. Containment policies have been the order of the day and this has meant that developers have built where they have been able to persuade planners to give them permission rather than in rationally planned and linked communities. Consequently, there has been a rather haphazard pattern of growth. Zoning policies have also meant that people have to travel long distances by car for work, shopping and leisure. The result has been a series of soulless and poorly designed housing estates on the fringes of towns and large employment estates dominated by ugly tin boxes near the main transport routes. This is not the utopia promised by planners 100 years ago. However, in spite of the fact that large parts of our post-war towns and cities look like a third-rate United States, planning has also had other impacts on the environment, notably in differences between areas. Thus these are now considered.

SELECTED STUDIES OF OUTCOMES AND IMPACTS ——————————————————————

Methodological issues

Chapter 3 began by stating that development control was hard to assess, because like birth control what hasn't happened can't be known. The very existence of development control has a 'deterrent effect' in that planning applications are not made if there isn't a reasonable chance of success. In addition there is the 'displacement' effect, which has affected behaviour by displacing activity (moving planning applications elsewhere). The 'deterrent' and 'displacement' effects make measuring the effects of development control a methodological nightmare.

Further problems are presented when an attempt is made to link development control decisions with plans and policies. Gilg and Kelly (1996a) have outlined seven difficulties in using development control data for assessing the efficacy of land use planning, four of which are germane to this section. First, policies are subject to change, leading to a time lag and overlapping of often contradictory policies. Second, policies can derive not just from different sections of the planning policy document arsenal but also from different public or private bodies, for example policies relating to infrastructure provision. Third, policies are only guidelines, even in the so-called plan-led 1990s, and can be interpreted by decision makers in many different ways. Councillors and planners are only human, with all the prejudices that this implies.

Fourth, policy makers can expect to achieve only a certain success rate. For example, Green Belt designation does not mean that no development will be allowed. However, few such policies contain explicit statements of what sort of permission rate is acceptable. So the suspicion remains that subliminally development controllers will adjust their permission rates to some implicit target like the apocryphal driving test examiner: pass one, fail one, pass one, fail one.

In spite of these difficulties much good work has been done with the data, and methodologies have been developed to circumvent them. For example, Brotherton (1992a, b) has attempted to use the concept of application quality, defined as the extent to which applications conform to the considerations that determine policy. Applications can then be categorised as weak or strong, depending on the degree to which they conform to policy, and similarly, decisions can be categorised as weak or strong, depending on the degree to which they confirm to policy.

It is important for applications to be weighted for their size and scale, since one application can be for 1 or 100 houses. In his study of the West Midlands Green Belt Gregory (1970) demonstrated this can make a considerable difference: he found that numerically applications for houses accounted for 50% of the total but that when weighted by area they accounted for 80% of the total. Thus all applications need to be weighted by house numbers/jobs created/area involved /local taxes created. The choice of weighting will depend on the type of land use being examined and the purpose of the study, for example, whether the study is concerned with land loss or job creation.

In a review of approaches to the analysis of development control decision making Gilg and Kelly (1996b) outlined four types of methodology: numerical analysis, a technical exercise, a power struggle and postmodernism, as shown in Table 4.9. The big division is between the first two types of analysis, which seek to 'number crunch' or 'eyeball' tables and maps of the way in which development control decisions vary over time and space in relation to plans and policies in order to assess the policy effect.

Table 4.9 The relationship between data sources and analytical methodologies

Data source	Numerical analysis	Technical exercise	Power struggle	Postmodern
Applications	•	•		
Appeals	•	•		
Observation of committee		•	•	•
Questionnaire of participants	•	•	•	
Interviews with participants		•	•	•
Case Studies	•	•	•	•

Numerical analysis refers to analysis by tables, figures and diagrams. *Technical exercise* refers to the procedures carried out. *Power struggle* refers to the struggle between key actors in the process to gain the ascendancy. *Postmodern* refers to the strong behavioural element in decision making and the lack of overall patterns.

Source: Gilg and Kelly (1996b)

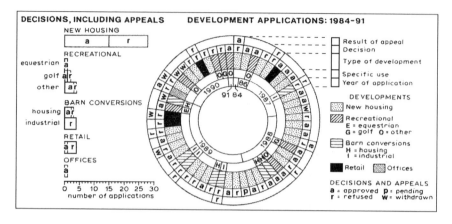

Figure 4.6 Wheel method of analysing planning applications. *Source*: Murdoch and Marsden (1994)

For example, Murdoch and Marsden (1994) have produced an innovative method of portraying planning applications as shown in Figure 4.6 based on a wheel system and a histogram. Although this shows only what happened in one village over seven years, it shows that the eyeballing approach has a lot to offer in a case study approach as both a summary device and one which should throw up interesting questions for in-depth study. Watkins *et al.* (2001) also employed new techniques, notably geographical information systems (GIS), to demonstrate that such studies can highlight significant variations in the patterns of the pressure to change land use and the responses of planners. Nonetheless, they caution that GIS can have significant resource costs both in terms of the system and in terms of staff training. However, although development control can be valuable the data do not explain why these patterns occur because they are 'end-state' statements of negotiative processes, which are often complex (McNamara and Healey, 1984) and thus need to be examined via case-study interviews of those involved in the negotiations. Examples of these approaches have largely been considered in the process analysis in Chapter 3, including Murdoch and Marsden's more detailed case-study work based on actor networks. In summary, development control data demonstrate the broad patterns of decisions made and where research should be directed if these patterns do not follow policy or show other aberrant patterns. The rest of this chapter outlines the main research findings classified by different types of environments and planning policies as well as the side effect of house price inflation.

Case studies of differential planning outcomes and impacts

National studies

In an attempt to relate decisions to plans Pountney and Kingsbury (1983) divided decisions according to their agreement or otherwise with the plan, as shown in Figure 4.7. The main finding was that there were few contradictions with the plan in any of the different areas. A second finding was that refusals rise with the rurality of the area, notably in Green Belts. However, plans are not always followed, because the plan may not contain specific enough policies, notably in market towns and Green Belts, where 'one-off' applications may be quite common or developers are 'testing' the resolve of planners to conserve these high-quality

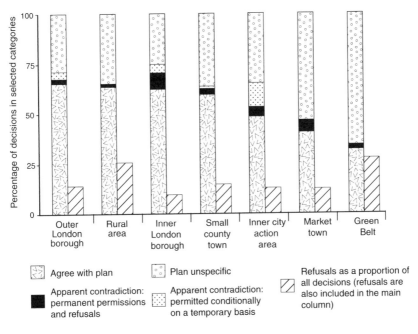

Figure 4.7 The relationship of decisions to Local Plans. *Source*: adapted from Pountney and Kingsbury (1983)

environments. Finally, the findings stressed the hybrid nature of decision making whereby each case is treated on its merits within the broad guidance provided by the plan. About 60% of the time plans were used, but at other times other factors came into the equation and the decision was 'plan-unspecific'.

Elson *et al.* (1998) in a major study of 3,490 development control decisions in 8 local authority areas found that approval rates varied by the area and the type of application. For example, in deep rural areas, approval rates were over 90%, in mixed rural areas they were 86%, but in the urban fringe they fell to 81%. By the type of application approval rates fell from over 90% for offices and garden centres to 86% for industry and tourism, to only 78% and 73% respectively for transport and retailing. The main reasons for refusal were inappropriate use in the countryside (31%), detrimental to residential amenity (27%), intrusion into rural landscape (21%), traffic issues (18%), design (17%) and poor access (13%) with obviously more than one reason sometimes being used. The main conditions were design (35%), landscaping (28%), parking (23%), level and hours of operation (14%) and access (10%).

Protected landscapes

It could be assumed that protected landscapes would attract fewer planning applications but Brotherton (1982) used aggregated data from the government to show that there were more applications the more rural and the more protected the landscape was, as shown in Table 4.10. This may seem a perverse response by applicants to policies designed to discourage planning applications, but the policy response was not perverse in that the refusal rate also rose in the same way. The perverse rate of applications can be explained by the desire of people to build houses and live in attractive areas and thus to apply against the odds on the basis of 'if you don't try you don't get'. However, Brotherton's work was severely criticised by McNamara and Healey (1984). First, for using unweighted data, and second, for using applications per thousand people as a measure of pressure. Third, they pointed out that the data reflect the end of a complex decision-making process and so did not show either the potential for planning officers to turn bad applications into good ones or the discretion available to planners in using policy or not when making decisions. In a response Brotherton (1984) argued that the data were not perfect but served to highlight areas where more detailed research into development control decisions should be directed.

Table 4.10 Development pressures in England and Wales

Area	No. of planning applications	No. of applications per 1,000 people	Refusal rate (%)
London	47,784	6.7	13.3
Metropolitan districts	78,571	6.7	14.2
England and Wales	484,267	9.8	16.3
Shire districts	358,854	11.7	17.3
National Parks	6,058	23.7	25.5

Source: Brotherton (1982)

In this vein, Curry and McNab (1986) as shown in Figure 4.8 demonstrated that rural areas attracted more applications per head of population than urban areas. However, part of this discrepancy can be attributed to applications coming from urban areas from would-be rural residents, with around 30% of applications coming from outside the immediate area. To some extent these patterns demonstrate the perverse effect of designation and countryside protection, since they make attractive areas even more attractive and sets up a vicious circle in which those fortunate enough to live in such areas are determined to protect them. Elsewhere development makes other areas less attractive and so the pressure on the protected areas is ratcheted up

Midgeley (2000) used a study of the planning register between 1996 and 1997 in North Yorkshire in which 147 applications for redundant building conversions were examined. The study area contained several different landscape designations, with some of them overlapping. As Figure 4.9 shows the approval ratio was lowest in the National Park (NYMNP), higher in the two district councils (HDC and RDC) but highest in the three AONB and AHLV (Area of High Landscape Value) areas, where 100% of applications were approved. This does not suggest a decisive impact by designated landscapes on decision making. Indeed, the analysis concluded that only in the National Park was the landscape designation the primary factor in decision making. Elsewhere planning policies in general were used, with landscape designation being a secondary factor. Crucially, there was more difference between LPAs than between landscape designation. A second piece of research asked planners to make a recommendation on a hypothetical planning application. Once again only in the National Park was the landscape designation the primary factor. Elsewhere there were differences between LPAs in their use of policy, with landscape designation being used as a back-up argument.

Finally, in Wales, Scott (2001) in a study of Special Landscape Areas in Ceredigion found

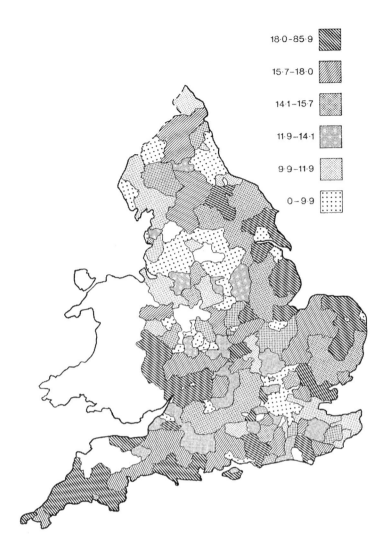

18·0–85·9

15·7–18·0

14·1–15·7

11·9–14·1

9·9–11·9

0–9·9

Figure 4.8 Planning applications per thousand population. *Source*: Curry and McNab (1986)

little difference between the rate of applications in the special areas and the rest of the study area. However, there was a greater desire to develop recreation and tourism in the special areas. Scott could not decide whether the special areas were deterring developers or whether countryside policies in general were too restrictive. He concluded that more research is needed and until then we cannot be sure what the impact of designated landscape policies has been.

Rural settlement and protected landscapes

Other studies have looked at the effect of two or more policies, notably the juxtaposition of settlement and landscape policies. For example, Gilg and Blacksell's (1977) work in Devon

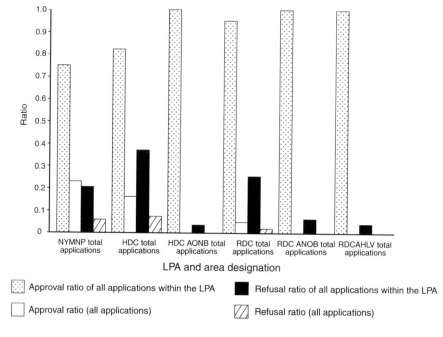

Figure 4.9   Approval and refusal ratios in North Yorkshire. *Source*: Midgeley (2000)

analysed around 20,000 planning applications in four contrasting areas of the county in an attempt to see if the pattern of permissions reflected policy designations in the relevant Development Plans. Clear patterns were identified but not always those expected. For example, in the south-east Devon study area as shown in Table 4.11(a) analysis focused on the effect of the AONB and settlement policy on development control decisions. The study area was divided so that half the area and half the population were covered by the AONB in order to see if the AONB deflected applications. Table 4.11(a) shows a clear deflection effect, with only 27% of applications being made in the AONB, although when residential applications are considered the ratio rises to 42% of applications. In this case the attraction of living in the AONB is encouraging a high rate of applications in spite of the policy designation. However, all other types of applications are deflected, notably commercial development.

Turning to the decisions made, Table 4.11(b) shows surprisingly that the rate of refusal is higher outside the AONB. But when the high number of approvals with conditions inside the AONB is considered this may be seen to reflect greater care with applications inside the AONB compared with more speculative and poor applications outside the AONB. When only residential applications are considered the refusal rates coincide, suggesting that AONB policy was not a crucial factor in the decision-making process. However, AONB policy is reflected in the greater number of conditions inside the AONB. Finally, Table 4.11(c) shows the effect of settlement policy in the AONB. At the time of the study Devon operated a key settlement policy, shown in Figure 4.10, under which about 70 settlements out of around 500 were designated for growth while all other settlements were supposed to take only minor

Table 4.11 The relationship between planning applications and Development Plan policies in south-east Devon (a) Planning applications by class of development

Class	Whole area	AONB section	% of total in AONB
All applications	2990	810	27
Residential	1305	549	42
Other buildings	961	204	21
Other developments	651	52	8
Commercial, etc.	73	5	7

Source: Gilg and Blacksell (1977)

Table 4.11 The relationship between planning applications and Development Plan policies in south-east Devon (b) Outcome of planning applications by type of landscape policy

Applications	Approval, no conditions		Approval with conditions		Refused		Total
	No.	%	No.	%	No.	%	
All	445	17	1408	55	725	28	2578
In AONB	102	15	452	66	129	19	683
Outside AONB	343	18	956	50	596	31	1895
Residential	105	9	612	53	431	38	1148
In AONB	16	3	291	59	187	38	494
Outside AONB	89	14	321	49	244	37	654

Source: Gilg and Blacksell (1977)
Note: The total number of applications in Table 4.11b is smaller than table 4.11a because 'Withdrawn' and 'Temporary' applications have been excluded.

Table 4.11 The relationship between planning applications and Development Plan policies in south-east Devon (c) Outcome by type of settlement policy

Village	No. of applications	Approvals	Refusals	Ratio
Newton Poppleford	126	68	64	1.06 : 1
Colaton Raleigh	55	42	13	3.23 : 1
East Budleigh	146	108	38	2.84 : 1
Otterton	68	46	22	2.09 : 1

Source: Gilg and Blacksell (1977)

developments. In the AONB only one settlement had key settlement status, Newton Poppleford (the first clear circle to the south-east of Exeter). But as the table shows this not only attracted the second highest number of applications, but also had the least successful permission rate of only 1.06 : 1, compared with ratios of over 2.09, 2.84 and even 3.23 in Colaton Raleigh.

Clearly, the policy was failing in two areas. First, it was not directing applications to the

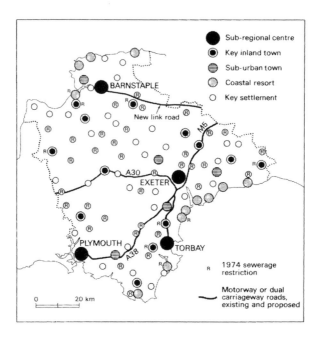

Figure 4.10   Key settlement policy for Devon. *Source*: Gilg (1978)

key settlements and decisions were less favourable to applicants in them. But of course the key
settlement and AONB policies are only two in a raft of policies, as shown in Figure 4.11, and
the wide diversity of policies makes it difficult to disentangle one policy effect from another.
Most notably, as Figure 4.10 shows, many key settlements were affected by sewerage
restrictions under which only a few dwellings could be built before existing sewerage facilities
reached their capacity. This demonstrates another planning failure, the ability to link policies
across different infrastructure providers.

Thus, other villages not chosen for growth but with potential for growth in terms of their
infrastructure, e.g. half empty schools or underused supplies of water, etc., grew dramatically,
as in the case of Clyst St Mary, a village just to the south-east of Exeter. As shown in Figure
4.12, Clyst St Mary grew enormously in the 20 years between 1945 and 1974, especially
between 1964 and 1974, when one estate provided 172 new houses. The village was attractive
to builders and house buyers because it was on an important road junction, only a mile from
a planned intersection on the soon to be built M5 motorway and only four miles from Exeter
city centre. But the estate was built on the other side of the busy main road from the primary
school and other facilities of the existing village and was built on top-quality land. In retrospect
the estate should not have been built, as it merely created a characterless suburban estate with
no relation to the existing village and increased commuting. In its defence it is well hidden
behind mature trees and makes little visual impact on the area, but it is nonetheless the sort
of development that should surely not be built.

However, the overall study of four study areas in Devon (Gilg and Blacksell, 1981)
concluded that the landscape and countryside protection policies appeared to have been
successful. Urban growth had been contained, and the built environment of the National Parks

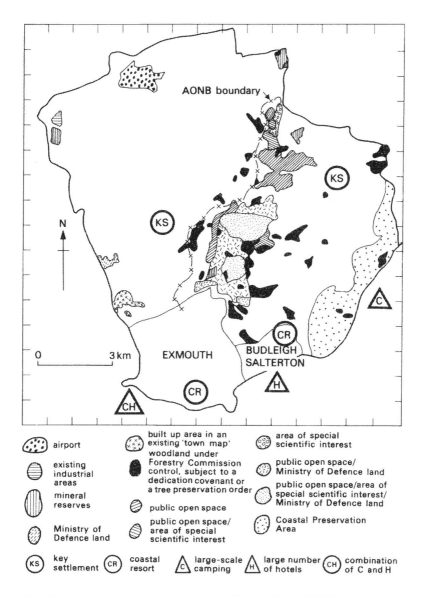

Figure 4.11 Planning constraints in east Devon. *Source*: Gilg and Blacksell (1981)

and AONBs had changed remarkably little. However, though development had been restricted to villages, it had not necessarily been restricted to the chosen villages. The overall pattern of decision making is shown in Table 4.12, from which it can be seen that the overall refusal rate was around 40%, double the government figure, because of weighting, but with significant differences between the ordinary countryside of mid-Devon (20% refusal) and the protected landscape of the Dartmoor National Park (55%). As in east Devon the most common decision is approval with conditions. In terms of the overall number of applications east Devon had

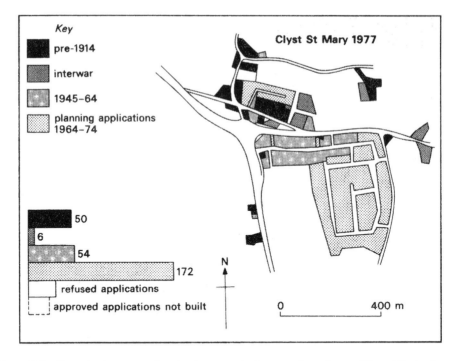

Figure 4.12 Example of an estate of modern houses tacked on to a village. *Source*: Gilg and Blacksell (1981)

eight applications per square kilometre compared with three in both south-east Dartmoor and mid-Devon, but only one in western Exmoor. Spatially, planning permissions were mainly kept within villages in east Devon, but elsewhere a significant number of applications were being allowed outside villages. In every area, though, key villages were not attracting the majority of permissions.

Table 4.12 Decisions on development control applications in four areas in Devon

Area	Approved, no conditions		Approved with conditions		Refused		Total
	No.	%	No.	%	No.	%	
East Devon	138	12.9	502	46.8	433	40.3	1073
South-east Dartmoor	42	5.8	286	39.5	396	54.7	724
Mid-Devon	22	6.7	240	73.2	66	20.1	328
West Exmoor	5	3.3	82	54.7	53	42.0	140
Total	207	9.1	1110	48.8	958	42.1	

Source: Gilg and Blacksell (1981)

Table 4.13 Land use changes and their relation to land use planning policies

Change	Area (ha)
Direct (compulsory) negative powers over change	
Farmland to developed land	129
Farmland to industrial land	39
Heathland to mining	22
Farmland to mining	16
Total	206
Indirect (voluntary) positive powers over change	
Heathland to woodland	126
Orchard to farmland	114
Farmland to woodland	95
Heathland to farmland	65
Total	400
No effective powers over change	
Farmland to unused land	88
Farmland to heathland	54
Total	142

Source: Gilg and Blacksell (1981)

The discrepancies can be accounted for as follows. In east Devon there was double pressure from the growing city of Exeter and second-home and retirement pressure to develop near the coast in the AONB. In south-east Dartmoor the traditional rural landscape and several large villages offer possibilities for infilling and farm-related development. In mid-Devon the presence of some villages and the weakest policies has attracted quite high levels of applications, but in western Exmoor an open moorland landscape and the presence of only two villages has clearly deterred applications, except in the adjoining north Devon coastal villages of Lynton and Lynmouth.

Finally, as Table 4.13 shows, land use planning has only a limited impact over landscape change. For example, only 206 ha of change went through the development control system, mainly farmland to developed land. In contrast, nearly double the total, 400 ha was transferred under indirect powers, mainly agricultural and woodland grants and subsidies, with notably 221 ha more woodland. When land transfers outside the system were added, Gilg and Blacksell found that nearly 550 ha of land had changed use outside the development control system, compared with 200 ha inside it.

Another study of all land use change relating to designated landscapes has been carried out by Pecol (1996), who used GIS derived from maps and air photos to study the relationship of policy to land use change in Bedfordshire. He found that areas with no designation had experienced change across 4.8% of the area, that land designated as an Agricultural Priority Area had experienced a 4.2% change, and that Green Belt land had experienced a 4.2% change. In AONBs and SSSIs, however, only 1.1% and 0.2% of land had experienced a change.

Anderson (1981) took several of Gilg and Blacksell's themes up in the early 1980s with work in East Sussex. First, she examined the whole county and found that applications per square kilometre varied between 10.1, 8.6, 7.0 and 5.1 in the four main towns of Hastings, Brighton, Hove and Eastbourne but fell to 1.7 in the town of Lewes and to around 1.1 in the rural parts of the county. This clearly demonstrated that planning applications were being directed not only to towns but to certain towns. As Table 4.14 shows there was also a clear spatial pattern, with some areas receiving six times more than their area would suggest (the coastal belt) but with other areas receiving three times less (the High Weald) and even six times less, as in the Sussex Downs AONB. Clearly planning policy was directing applications to some areas rather than others.

Table 4.14   Planning applications, by area, in East Sussex

Area	All applications		Approximate area	
	No.	%	Km^2	%
Coastal urban belt and Lewes	1843	57.5	160	8.9
High Weald (rural areas)	613	19.1	870	48.3
Low Weald	506	15.8	520	28.9
High Weald (urban areas)	176	5.5	10	0.6
Sussex Downs AONB	65	2.0	240	13.3
Total	3203	100.0	1800	100.0

Source: Anderson (1981)

She then divided the rural part of the county into three areas: the Sussex Downs AONB, the proposed High Weald AONB and average countryside, the Low Weald, as shown in Figure 4.13, to see if any policy effect could be seen in these three rural areas. As can be seen in Figure 4.13 a clear pattern for permitted applications was demonstrated. In particular, there were very few applications in the Sussex Downs AONB. There was little difference between the other two areas, with a few more applications in the Low Weald, notably near the coast and in the urban fringes of Eastbourne and Hastings. The location of refused applications as shown in the lower half of Figure 4.13 reflects quite closely the pattern of permissions. This suggests that once the policy effect has been seen in the pattern of applications it is not repeated in the pattern of decisions, which are based on the merits of each case, so that a sound application in the AONB has as good a chance as elsewhere.

Anderson then compared her results with those obtained by Gilg and Blacksell and concluded in general that key settlement and AONB policy had been more effectively implemented in East Sussex than in Devon, perhaps reflecting the longer time that policies had been in operation in East Sussex.

Suburbs and conservation areas

Whitehand and Carr (1999) examined development control in the suburbs of London and Birmingham between 1948 and 1993. They found that the number of new dwellings had fallen markedly over the period, with to some extent extensions taking over in terms of numbers. For example, in 1960 each area had around 25,000 applications for new dwellings but by 1990 this had fallen to 15,000 in London and only 5,000 in Birmingham. In contrast, applications for extensions had risen from virtually nothing in 1960 to around 1.5 extensions per 100 dwellings. Both these figures imply that the suburbs may not be the potential sites for substantial infilling on 'brownfield land' desired by the government and implied by the

Permitted

One dot represents 1 or 2 applications
to a total of 5 dots in any km^2

▨ 11+ applications in any km^2

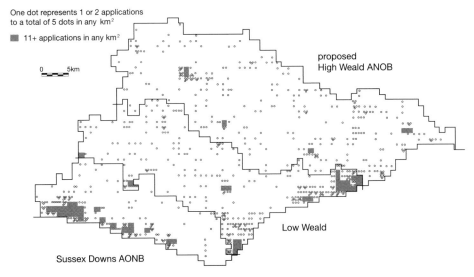

proposed
High Weald ANOB

0   5km

Sussex Downs AONB

Low Weald

Refused

One cross represents 1 or 2 applications
to a total of 5 crosses in any km^2

▨ 11+ applications in any km^2

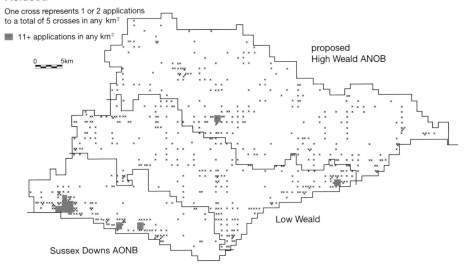

proposed
High Weald ANOB

0   5km

Sussex Downs AONB

Low Weald

Figure 4.13   Location of permitted and refused planning applications in East Sussex, 1975–76. *Source:*
adapted from Anderson (1981)

research work of Best and Sinclair. Indeed, Whitehand and Carr found only a weak
correlation between density and applications for new dwellings. Ironically, there was more
demand in high-density areas and low-density areas. This suggests that high-density areas are
attractive for development in terms of student or retirement accommodation and that low-
density areas offer potential for either new apartment blocks or one house.

In a further study of Edwardian suburbs in Birmingham, Whitehand and Morton (2003) found that they were not subject to specific policies but were subject to haphazard and often strong pressure for redevelopment, often because of the substantial size of the buildings or their large gardens. Allotments and large public buildings were also key targets. However, planning decisions had contributed to the suburbs retaining their essential character. In contrast, most suburban areas after the Edwardian era are of average density and are thus unsuitable for infill because the gardens are too small, any new developments would be very incongruent in the landscape and would be strongly resisted as a threat to amenities and property prices. In addition, as Adams *et al.* (2001) have shown, the distinctiveness of land as a commodity, the imperfect nature of the land market and the behavioural characteristics of landowners pose considerable constraints on the amount of brownfield land that will actually be available for development at any one time. Thus success rates of getting more than 60% of new houses built may not be sustainable and are probably due as much to the property slowdown of the late 1990s as to planning policies and also to how people want to live.

In the special case where a suburban or inner urban area has distinctive character and has a lot of listed buildings or is a Conservation Area, Larkham (1992) has found that interest in listed buildings as measured by membership of amenity societies is very much restricted to a few favoured settlements. For example, Hampton in Arden, near Stratford on Avon and full of Shakespearean heritage, has a very high membership of 34.5% of the population. However, even middle-class Moseley and fashionable small town Ludlow have only around 10% of their population as members. Most settlements, though, have less than 1% of their population as members, including surprisingly the market town of Shrewsbury. Thus though the number of amenity societies grew from only 200 in 1957 to around 1,000 by the late 1980s, this does not necessarily reflect mass interest in conservation of listed buildings.

Table 4.15 Representations on planning application to convert a cinema to a retirement residence

Comment	No. of responses
Retain cinema for arts and community purposes	1,114
Site on a busy road junction not suitable for old people	323
Access and sight lines poor	318
Parking provision inadequate	317
Site too noisy	104
More consultation time needed	6
Housing needs for elderly catered for elsewhere	5
Housing could be built without demolishing cinema	2
Cinema is example of 1930s' architecture	2
Four-storey building is out of keeping	1
Too many old people in Ludlow already	1

Some 1,119 individuals and 9 organisations made comments.

Source: Larkham (1992)

Most interest is found to occur when a neighbourhood is threatened by change, and Larkham (1992) demonstrated that virtually all properties around a planning application to convert a cinema to a retirement residence objected to it. As Table 4.15 shows, comments centred on three themes. First, the desire to retain the cinema, not necessarily a land use issue. Second, traffic issues, definitely a land use issue. Third, noise, something which planners can do little about. Significantly there were virtually no comments about the character of the building and the area and so it can be argued that people care little about their built environment, even when it is threatened with change.

In terms of the type of changes requested Larkham found that the distribution was fairly even. However, applications for conversion of outbuildings to residential use were by far the most common, followed by conversion of single dwellings to flats. There were disappointingly few applications to convert upper floors of shops to residential use. Larkham suggests that the Conservation Area is attracting considerable interest from developers seeking to provide small housing units for presumably young and old people, which should help to safeguard the vitality of the town centre. Indeed, since Larkham conducted his study Ludlow has become a fashionable weekend resort for people attracted to its 'foodie' restaurants and old-world small town ambience.

House prices

One of the most unfortunate side effects of restrictive planning policies has been excessive house price inflation, as the supply of building land and thus new houses has been severely restricted. Although the media welcome house prices as good news since existing house owners appear to benefit from an increase in what is normally their biggest capital investment, house price inflation is a bad thing. First, because inflation is *per se* a bad thing and other price rises in the rest of the economy are definitely not welcome. Second, increasing house prices may benefit people marginally when they 'trade down' in retirement to a smaller, cheaper house and so can use the surplus cash for other things, but for most of the time the cash is locked up and is not being spent in the wider economy. Third, high house prices are unfair to young people trying to get on to the ladder or for those people where retirement or second-home migration forces prices up and so forces local people off or down the property ladder. Nonetheless, the relationship between planning policies, housing demand and supply is a complex and contested one, as the discussion below will demonstrate.

The percentage of land prices to house prices has been rising constantly since the 1960s. As already outlined, Peter Hall's classic study showed that land prices accounted for 10–25% of new house costs in the 1970s. By the early 1990s land represented over 34% of the cost of a house (Meikle *et al.*, 1991). By 1994 land had risen to as much as 46% of the price of a new house in areas like Basingstoke (Campbell *et al.*, 2000). Furthermore, Campbell *et al.* noted that new costs like planning obligations which impose on developers the cost of providing both on-site and off-site infrastructure had meant that the construction cost of a house was less than a quarter of the price paid by the purchaser in the 1990s. In other words the bricks and mortar and plumbing, etc., account for only 25% of the price of a new house.

This inflationary impact of constraining the supply of land for houses also continued, so that by the year 2000 it was reported that since 1970 house prices had risen in real terms by 135% compared with a real rise in wages of only 60% (*Planning*, 8 December 2000). Indeed, as Table 4.16 shows, house prices by 2002 were on average seven times higher than average earnings. The inflationary impact of planning constraints is also shown with the South East, with its restrictive policies having house prices eight times over average earnings. The regional pattern is, however, also closely related to pay, except in the South West, which attracts much second-home and retirement house buying. Ironically, the relatively more affordable house prices in the North should in theory encourage people to move there and take some of the pressure off the South, especially as the county data show that there are some dramatic variations within the regional averages. These very low prices, however, indicate very run-down, crime-ridden and deprived areas where in some districts houses are virtually unsellable. Planning has not been able to solve the long-standing north–south divide and may even have exacerbated it by dividing Britain into two nations, those who cannot afford to move to the south and those who dare not leave the south.

Table 4.16 House prices related to average wages, by region (£)

Region	Average weekly pay	Average house price as % of annual pay	Average county price
London	497	888	London 249,000
South East	352	851	Hertfordshire 194,000
South West	296	803	Devon 120,000
East of England	327	748	Norfolk 111,000
West Midlands	308	598	Warwickshire 124,000
East Midlands	292	568	Leicestershire 102,000
Yorks + Humberside	294	488	West Yorkshire 81,000
North West	310	483	Cheshire 116,000
North East	286	459	County Durham 57,000
England	339	702	

Source: *Regeneration and Renewal*, 26 July 2002, p. 10; *The Times*, 1 July 2002, p. 4, for county prices

Bramley (1993) has attempted to link planning policy and house prices via a demand and supply model. The model suggested that the impact of the release of much more housing land would not be very big and there would be only modest benefits. The release of Green Belt land would, however, make some difference, notably in urban fringe areas. Infrastructure charges on house builders would lower their output and would be reflected in the cost being passed on in the form of higher prices. Finally, rising house prices only slightly encourage builders to apply for planning permission. Bramley's work was, however, theoretical and was carried out at a time of negative equity in the housing market. This was caused by the cyclical nature of the housing market, which reflects boom–bust cycles in the wider economy. Because housing is such a key factor in people's aspirations not just as a place to live but also as their major capital investment, the government uses interest rates and thus mortgage rates as a key tool for managing the rate of growth in the economy. When interest rates are low people tend to borrow more than they can really afford, given the history of high inflation, which for long rendered over-borrowing sensible, because high inflation rapidly reduces the debt. In the early 1990s, following such a boom, interest rates rose and house prices fell so that people found they had more debt than the house was worth, hence negative equity. This period demonstrates another facet of the house price phenomena.

In contrast to Bramley's theoretical work Monk and Whitehead (1999) attempted to measure the actual effect of planning on house prices in three districts in Hertfordshire and Cambridgeshire. The basic hypothesis was that planning permission would be easier to get and house prices would be lower the farther one travelled northward away from the overheated market in London. The first evidence, as shown in Figure 4.14, was that stocks of planning permission did follow the expected trend, with north Hertfordshire always having the lowest stock and Fenland having the highest stock in the second half of the study period, 1988–91. In this period a house price explosion was followed by a severe collapse in prices

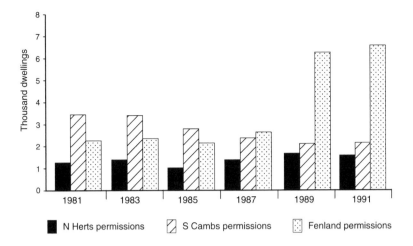

Figure 4.14 Stocks of planning permission, 1981–91. *Source*: adapted from Monk and Whitehead (1999)

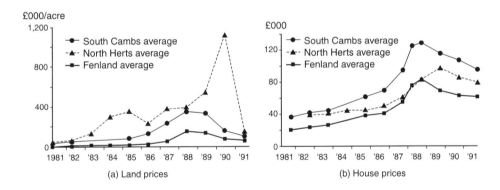

Figure 4.15 (a) Housing land and (b) house prices, 1981–91. *Source*: adapted from Monk and Whitehead (1999)

and the 'negative equity' of the early 1990s in which people owed more on their mortgage than the house was worth. In less pressured areas like Fenland houses became difficult to sell and speculative planning permissions acquired in the boom of the late 1980s were not taken up. This boom-and-bust cycle is shown in Figure 4.15(a), where north Hertfordshire can be seen, as hypothesised, to be the area where land prices rose most dramatically in the late 1980s boom. However, as shown in Figure 4.15(b), north Hertfordshire did not have either the highest house prices or the highest house price rise, which may be explained by the booming economy of Cambridge at the time. It also suggests that the link between planning and house prices may not be a linear one, because the sheer number of existing houses in an area are out of all proportion to the small number of houses coming through the planning system in any one year.

The most crucial evidence, however, came from the relationship between plan forecasts and house building. Starting in north Hertfordshire, as shown in Figure 4.16(a), it can be seen that though the provision was reduced during the period, completions were above the forecasts for nearly every year except the recession years of 1981–82 and 1991. In south Cambridgeshire, as shown in Figure 4.16(b), the opposite occurred, with forecast completions increased, but in only two years were these forecasts exceeded by housing completions. In Fenland, where planners wanted to attract development, forecasts were radically increased and house builders responded well to the signals by exceeding the forecasts in virtually every year, as shown in Figure 4.16(c). However, development 'fever' in Fenland was overdone and as Figure 4.14 shows it remained the area where most planning permissions were unused and with the lowest house prices. In other words, Fenland was attractive in boom times but not in normal times, and its growth may have been due to the expansion of Peterborough as a sub-regional site for London overspill rather than from direct overspill emanating in waves from London through north Hertfordshire and Cambridgeshire.

The evidence appears to confirm the hypothesis that planning policies become more relaxed as one moves away from a pressured area with the exception perhaps of Cambridge, which anyway is a special case because of a thriving economy based on its university and spin-off services. Monk and Whitehead concluded with two main findings. First, planning raises prices, promotes speculation and exacerbates the boom–bust cycle in housing which is related to the wider economic cycle, notably interest rates. Second, restraint in one area is reflected by displaced pressure next door but not enough to meet overall demand, as in Cambridge, but such pressure is not severe enough to sustain demand in more peripheral areas.

Monk and Whitehead's work has two main implications for planning. First, planning cannot be blamed alone for high house prices, but it is a contributory factor, especially when severe restraint policies are followed. Second, restraint policies have only a limited effect in encouraging people and jobs to move to less restrained areas, if the conditions favourable to them are not replicated in the less restrained areas. Thus planning policies have to be proactive in creating the right conditions for employers and residents. Internationally, research by Dawkins and Nelson (2002) has found that the demand side of the housing market may be a stronger determinant of prices than local urban containment policies. Nonetheless, they conclude that local planners can play a significant role in relation to the severity of house price inflation attributable to urban containment policies.

In a related study Crook and Whitehead (2002) examined how planning policies might provide affordable housing for those unable otherwise to buy in the 'free market'. They argued that most local planning authorities would respond positively to a central government requirement to have an affordable housing policy. However, the number of houses built would depend on their will to implement the policies, their negotiating skills and the market in their area. In areas of low demand and high costs, such as inner urban areas in the North, the potential to achieve affordable housing is likely to be limited. In contrast, in areas of the buoyant South the potential could be great, especially on greenfield sites.

Finally, in December 2003 Kate Barker, a member of the Bank of England's monetary policy committee, issued a preliminary report on the link between planning policies and house prices to the Chancellor of the Exchequer, Gordon Brown. She identified land supply as the root cause of the problem, with house builders having only about one year's supply of land with planning permission. This is because the planning system releases too little land, too slowly, largely because there is little incentive for local councils to grant more planning

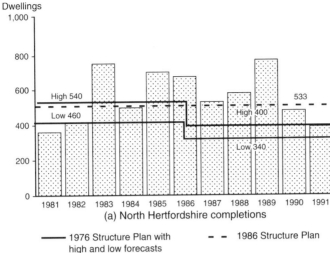

(a) North Hertfordshire completions

——— 1976 Structure Plan with high and low forecasts   – – 1986 Structure Plan

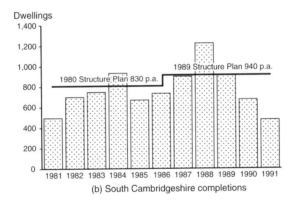

(b) South Cambridgeshire completions

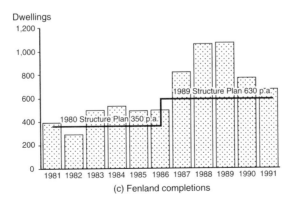

(c) Fenland completions

Figure 4.16   New housing completions in (a) north Hertfordshire, (b) south Cambridgeshire and (c) Fenland, 1981–91. *Source*: adapted from Monk and Whitehead (1999)

permissions. First, because new development is costly in terms of infrastructure and second, because it is unpopular with local residents. She concluded that an extra 39,000 new houses a year were needed to accommodate the current population and an extra 145,000 a year would be needed to bring English house price inflation into line with Europe. These figures compare with recent rates of only 140,000 new houses a year in England and 155,00 in the United Kingdom. Until this was achieved the cost of current policies would be £8 billion sucked out of the economy and first-time buyers paying £32,000 more than they should be. In addition only 37% of young people can now afford to buy a house, compared with 47% in the late 1980s.

In her final report in March 2004 Barker (2004) made the following recommendations to achieve four objectives: first, improvements in affordability, second, a more stable market, third, housing locations which support patterns of economic development, and fourth, an adequate supply of publicly funded housing. She set out three scenarios. First, to reduce real house price inflation to 1.8% (compared with a long-term rise of 2.4%) an additional 70,000 private houses should be built. Second, to reduce inflation to 1.1% an additional 120,000 houses should be built. Third, to meet the need for new households an increase of 17,000 social houses should be built or 26,000 extra houses if inroads were to be made into the backlog of need. She also recommended that planners should take more account of and make more use of market information and thus allocate more land for development. The government responded by considering how regional planning could manage regional housing markets better and by urging local authorities to be more responsive to local housing markets, but still retained the target of building 60% of houses on brownfield land. However, Entec *et al.* (2004) estimated that the Barker recommendations would take 77,500 ha of greenfield land by 2016 (half of Greater London). They would also lead to a 20% increase in carbon dioxide emissions, 9.3 million tonnes of extra waste and a demand for an extra 50,000 megalitres of water. Other estimates by the CPRE (2004) forecast a total greenfield loss of 1.18 million ha, or 26 Sloughs.

Thus we see in this report a clear indication that land use planning has had a dramatic impact on the social and economic life of Britain, with winners and losers. Winners are retirees or home owners trading down, property speculators, outward migrants and landowners. Losers are first-time buyers, home owners trading up, inward migrants and other non-homeowners. On the wider stage, one of the reasons why Britain could not join the euro in 2003 was said to be the difference in house price markets between the two areas. This is partly due to different types of interest rates (long-term in Europe) and different attitudes to home ownership. (In the United Kingdom a house is seen as an investment, in Europe as a place to live/rent.) But Britain is the second most densely populated country in Europe, and as the Barker report also concluded new houses have been built shoddily, using outdated techniques, and so the solution surely cannot be more rabbit hutches on postage stamps on isolated estates far from jobs and shops. Some of the possible solutions are discussed in Chapter 5 but before this it is time to discuss the key issues in this chapter.

Discussion and summary

As we have just seen, policies have been dominated by restraint. In an analysis of the first generation of Structure Plans Cloke and Shaw (1983) have demonstrated, as shown in Figure 4.17 that severe restraint policies were a feature of most of the plans around London. Elsewhere most counties operated restraint via key settlement approaches which restricted

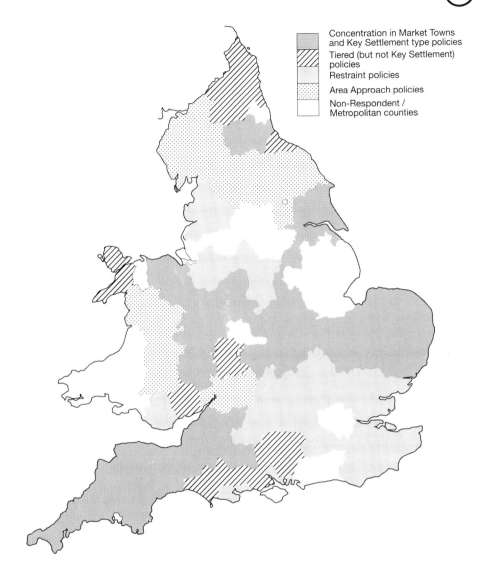

Figure 4.17 Settlement policies in Structure Plans. *Source*: adapted from Cloke and Shaw (1983)

growth to typically around 10% of villages and market towns, in order to concentrate services and get economies of scale and to conserve the open countryside. Murdoch *et al.* (1999) have shown how a few people have continued via middle-class groups to create an anti-development ethos in Buckinghamshire. This chapter has shown, however, that much growth has taken place in these areas and this fact opens up the argument as to whether planning can buck trends or merely guide them.

Elson (1981) has argued that restraint policies had to give way to economic pragmatism, a view endorsed by Cross and Bristow (1983), who have argued that planning has become an incremental process arbitrating between powerful interests and specific issues and has lost any pretensions to being a long-term or comprehensive decision-taking process. Cloke and Shaw (1983) agree and have concluded that the planning policies of the authority are intentions, desirable courses of action and merely a baseline for negotiation.

Although most of these observations were made in the 1980s and referred to plan making, the research results outlined above confirm that planning policies are only guides to action on the ground and that the key element in what happens is the negotiations between the key actors involved. There is only a weakish link between planning outputs (plans, policies and development control decisions) and planning outcomes and impacts. It is known as the policy implementation gap (Gilg and Kelly, 1997b). The gap is not necessarily a bad thing, because planning policies in Britain are deliberately flexible and each planning application is treated on its merits. Nonetheless, the degree to which plans are used in assessing the merits of each case is a key debating point – especially since the greater emphasis given to plan making in the 1990s, with first the so-called plan-led system in the early 1990s and then the introduction of more prescriptive Regional Plans in the late 1990s.

The link between plans and development control lies at the core of planning and notably any evaluation exercise, but in isolation it may be misleading. Accordingly, Chapter 5 attempts a wider evaluation of planning although keeping the policy implementation gap firmly in focus.

Further reading
in addition to references already highlighted in text

Cheshire, P. and Sheppard, S. (1989) British planning policy and access to housing, *Urban Studies*, 26, pp. 469–85.

Curry, N. (1992) Controlling development in the National Parks of England and Wales, *Town Planning Review*, 63, pp. 107–21.

Department of the Environment (1991) *Rates of Urbanisation in England 1981–2001*, by P. Bibby and J. Shepherd, London, HMSO.

Department of the Environment (1991) *Housing Land Availability: A Study by Roger Tym and Partners*, London, HMSO.

Department of the Environment (1993) *Housing Land Availability: the Analysis of PS3 Statistics on Land with Outstanding Planning Permission*, by P. Bibby and J. Shepherd, London, HMSO.

Foley, P. and Hutchinson, J. (1994) Planning applications as a primary data source: industrial development in Sheffield 1978–1991, *Area*, 26, pp. 123–32.

Most journals are available electronically and search engines will trawl the net for research papers and monographs but beware that some sites may not be subject to refereeing and may therefore contain polemical information. The continuing saga of the housing debate can be followed on the 'Sustainable Communities' Web site.

5

A Set of Evaluations and the Way Forward

The main purpose of this chapter is to provide an evaluation of the planning system and to assess proposals for change. There are two types of evaluation. First, those based on specific approaches, notably inferential assessment, historical evolution, typology and theory. Second, a hybrid approach which combines the evaluations throughout the book to evaluate key themes grouped under seven checklist headings based on the Morrison and Pearce assessment model discussed in Chapter 2.

Learning objectives

1 To consolidate the understanding achieved in Chapters 1–4.
2 Then to go from understanding to explaining the system by developing rigorous and theoretically informed explanations of the system.
3 To reflect on the system, evaluate it and make proposals for change.

REMINDER OF THE ISSUES AND PROBLEMS IN EVALUATING PLANNING

The system is hard to evaluate because of the 'what if' problem or the 'birth control' and 'displacement' effects of planning. These 'birth control' and 'displacement effects' will always be hard to measure, but as Pickering (1987) reminds us with birth control the real effect is measured by what doesn't happen. Pearce (1992) has added four further problems, which overlap with the fuller discussion of similar problems outlined by Gilg and Kelly in earlier chapters. First, whose goals should be evaluated? Second, official goals are difficult to trace and they have changed over time. Third, there are often unintended consequences. Fourth, there are limited data and statistical methods to compare goals with results. Another problem is that planning is not a 'free agent' and is constrained by the context in which it operates. For example, planning policy is not contained in one core document but in several documents, and policies are delivered by different organisations and mechanisms. This would be bad enough but the mixture is only one of many agencies (stakeholders) that have an impact on what planning can and cannot do. Planning can thus by and large only react to the actions of other agencies, though it does have one crucial power, the ability to say *no*. Finally, planning is constrained by the planning constraints coffin shown in Figure 2.5.

Nonetheless, as earlier chapters have shown, evaluations have been made, but Preece (1990), in an excellent overview of evaluation methods, is critical of two frequently used methods. First, the common procedure of comparing one area with another, which, he argues, is simplistic compared with say field trials for new crops, which divide fields into at least 10 plots. Second, there is the fallacy of affirming the consequent, in which A is hypothesised as having an effect leading to B, and if B is observed the hypothesis is said to be true. But this is fallacious. For example, if before entering a railway station we asserted that throwing paper out of the carriage window would lead to a lack of elephants on the platform, we might well believe the hypothesis to be true when no elephants were observed at not just the next but all stations en route. Another problem with the fundamental fallibility of using past evidence to predict the future is that there is no guarantee that however many times one event has been followed by another it will always be so. However, the fact that past evidence cannot 100% predict the future does not mean that the method is 100% unreliable. True, there's no logical guarantee that the future will resemble the past, but there are good grounds for supposing that it will.

As an alternative to these two common errors, Preece argues, quantitative and qualitative methods should be seen as complementary, and replication and differentiation should be the main testing methods. Replication should involve the comparison not just of two areas but of several areas, and they should be chosen not at random but because of their different characteristics. Differentiation should be used to observe that policies were observed to change differences between areas, even if this ironically could be a decrease of difference, as for example, the aim of inner-city regeneration. Finally, a time period of at least 10 years should be chosen in order to smooth out any aberrations.

In conclusion, Preece advocates that work (1) should be logically rigorous, (2) should be based on proper hypothesis formulation and (3) should investigate positive and negative policy objectives equally. In general it should also follow the advice of the historian Hugh Trevor Roper, that history (land use planning evaluation) is not just about what happened but also about what happened in the context of what *might* have happened. Preece's advocacy is, however, hard to follow when evaluating planning, and few if any studies have come close to meeting his strict criteria. In this sense, evaluation has a long way to go and the following sections will highlight how few studies meet Preece's criteria, since most of them are based on descriptive assessments, historical evolution and checklists rather than rigorous theorisation. Nonetheless, however imperfect the methods, evaluations can be made and they should focus on seeking to find causal links between objectives, outcomes and impacts via the outputs achieved as outlined in Chapter 2.

FOUR TYPES OF EVALUATION

As Chapter 2 set out, four main types of evaluation can be found in the literature: inferential qualitative assessments, evaluation by historical evolution, evaluation by typology and evaluation by concept and theory.

Inferential qualitative assessments

These are assessments made on the basis of experience and wide reading of the literature. In this genre Rydin (1993, 1998) has made two simple but important observations. First, the strength of the system rests on its attention to detail and treating each case on its merits, but second, the failure of the system is in not seeing the wider picture, not assessing the impact that it is having and in being too reactive. This is largely due to the split between plan making and development control and the lack of monitoring that occurs within local authorities. This was highlighted by the Audit Commission (1992) in its report on development control that found:

> In practice most authorities lack systems to ensure that their systems are faithfully implemented . . . most planning departments, having given attention to the decision-making process, do little to monitor and thereby demonstrate the success of development and their achievements in the quality of the outcomes.
>
> (p. 40)

Nonetheless Pickering (1987) has concluded that development control has played a crucial role in producing the quality of environment we see around us in Britain today. Its most spectacular success has been in maintaining, in most places, the clear division between town and country.

In a similarly congratulatory mode Diamond (1995) has been able to credit post-war planning with five successes and only two failures. The five successes are: first, achieving significant improvements in urban form; second, achieving the virtual abolition of slum housing; third, effective management of land use conflicts; fourth, undertaking countryside planning for city dwellers; fifth, considerable success in containing regional disparities. The two failures have been: first, limited examples of successful inner-city regeneration; and a failure to effectively relate transport and land use planning.

Pearce (1992) has provided an excellent summary. He begins by providing a list of the post-war goals of planning, which are generally agreed to have been:

1   Facilitating and encouraging private enterprise.
2   Ensuring the supply of land for private housing.
3   Containing urban areas.
4   Reusing derelict land.
5   Protecting residential amenities.
6   Conserving historic or heritage buildings, and the landscape.

The list used to be longer but in recent years some previous goals have been downgraded, notably attempts to overcome regional imbalances by redistribution; planned decentralisation; securing self-contained and balanced communities; public expenditure on the provision of public housing and new settlements; and segregating employment use and residential use.

The main achievements have been: first, protection of countryside and residential areas; second, increased reuse of derelict land; and third, more efficient use of public infrastructure. In contrast, the main unachieved goals or unintended costs have been: first, reduced economic growth due to delay plus constraints and bureaucracy; second, inflation of land values; and third, separation of home and workplace.

The possible sources of failure have been: first, information imperfections; second, bureaucracy and lack of co-ordination; third, the adoption of multiple and conflicting goals; fourth, lack of competition between monopoly suppliers of planning services; fifth, limited administrative capacity and resources; and sixth, government externalities, for example, 'public bads' located in marginal areas.

Cullingworth (1994, 1996a, b, 1997, 1999) in a series of reviews of 50 years of British planning, argues that four key themes stand out:

1   The two key philosophies of containment and redistribution have remained central.
2   Cross-country commuting has grown as houses have leapfrogged Green Belts.
3   Change is seen as something to be resisted rather than facilitated, notably new building in the countryside.
4   Policies allow applicants to cherry-pick a policy to suit them.

Cullingworth also argues that planning has failed to cope with change because:

1   Government has tried to mediate between conflicting interests rather than leading.
2   Transport and land use planning have been separated, and the growing separation of the places of work, residence and recreation has created an insoluble transport problem that is a tragic example of postponing the necessary action.
3   Lack of economic analysis in planning.
4   The system is too adversarial.

Davies (1998), also in a review of the first 50 years of planning, has argued that the five core principles of the 1947 Act remain: the definition of development; the duty of LPAs to prepare plans; planning permission is related to the Development Plan; the right of appeal; and the power of enforcement. However, the practice of planning has radically altered. In particular, planning is no longer based on public welfare provision but works with market forces. Planning is no longer based on monolithic local government but has to engage with many public bodies and quangos. Planning is no longer dominated by local issues but is influenced by European and global forces. Nonetheless, planning has had two great successes: containment and conservation.

Thomas (1997) has argued that planning has been good for some people, notably those who have gained permission to develop, but less good for those affected by major developments near their house. In particular he notes that McCarthy's (1995) survey of public attitudes found that people thought that developers had benefited at the expense of individuals. Nonetheless, most people thought that the planning system was essential in a small crowded island but that it could be operated better and could be less adversarial. Thomas also criticised the cost of the system, in terms of delays and the inflationary effect on land values. In particular he criticised the way in which betterment, though created locally, is creamed off by national taxation. Instead, betterment should be retained by the local community. Other criticisms by Thomas focused on weak public participation, potential officer–member conflicts, and poor attention to equality issues, notably in terms of gender, disabled people and ethnic minorities.

Finally, Ward (2004) has provided an evaluation of planning impacts since 1945 by the type of area and type of impact. In *city centres*, Ward argues, planners have preserved their commercial function by facilitating redevelopment and encouraging alternative uses like tourism where decline has been a problem, notably in places like Liverpool and Glasgow. In the *inner city*, Ward posits, the planners' role was minor, faced with structural decline in these areas, as their environment and labour force were unsuited to the new service economy. In recent years regeneration policies have had some impact, but not really to the benefit of the now mainly immigrant and less skilled work forces. In the *suburbs* there has been little change except for infilling, but there has been a slow decline in quality of life as suburban shopping centres have not been able to compete. In addition, the general lack of work and leisure facilities in areas dominated by housing has contributed to traffic problems.

In the *outer city* decentralisation has taken place but not in the planned way envisaged in the 1950s, with public new towns being replaced by private estates tacked on to existing towns and villages beyond the Green Belts. Ward argues that this process has been planned only in local detail, but that spatially developers have opportunistically sought and found gaps and weak points in planning policies beyond the Green Belt. Planners have guided the process, but the key drivers have been private developers, mortgage lenders and the private car. This has produced a piecemeal pattern of post-war development and the spatial incongruence of homes, jobs and shops has necessitated more daily travel within what has become the dispersed city. Ironically, the general success of containment policies has preserved most of the countryside in the outer city, thus allowing the expansion of the road infrastructure which carries most of these daily road movements, for example the M25 around London and the M42 around Birmingham.

Regionally Ward suggests that planning promoted social stability at the cost of economic growth by attempting to divert growth away from the South East between the 1950s and the 1970s, but that since then planning has promoted economic growth at the expense of social stability. Where regional dispersion has taken place it has favoured leafy green cities like Norwich or Exeter or suburban locations around the big cities of the North. The depressed inner cores of the major industrial cities have not benefited from regional policies. *Economically* Ward argues that the biggest planning impact has been the restriction on the supply of building land and the concomitant dramatic rise in land values, which may have depressed GDP by as much as 10%. Builders have, however, gained from the near certainty that planning permission has provided a sellers' market and thus reduced the risk in house building.

Finally, Ward argues that the *social impact* of planning has varied between social groups. Older owner-occupiers have done well, but younger owners have had to face dramatic swings in house prices and new houses, which are little better than rabbit hutches on postage stamps miles from jobs and services. Public-sector tenants have done well in the new and expanded towns and those who have bought their properties on some of the better council estates. But a significant minority now live on 'sink' estates with severe social problems of low employment and high rates of crime and drug taking. Other losers have by and large been ethnic communities. The changing role of women in the home and the workplace has not by and large been recognised, with traditional suburban homes far from jobs and shops, exacerbating the problem of combining work with child care for many mothers.

Overall, Ward concludes that planning has benefited those who would have done well without planning but harmed those who have been forced to live in over-planned housing redevelopments. But for most of the time and for most people planners have 'blandly adjusted a pattern of urban change that has been shaped by powerful economic, social and political processes that were largely beyond their control' (p. 285).

Evaluation by historical evolution

An evolutionary approach is useful because British legislation, as pointed out in Chapter 2 is incremental and based on precedent. Healey (1988) has used the concept of the Gilg–Selman spectrum to show how the main form of regulation (development control) did not alter at all, while other policy measures have been modified on average every 10–15 years. Planning has been subject to the vagaries of fashion and paradigm shifts in both society and academia. In terms of design the post-war period has seen two main periods, as shown in Figure 5.1. First,

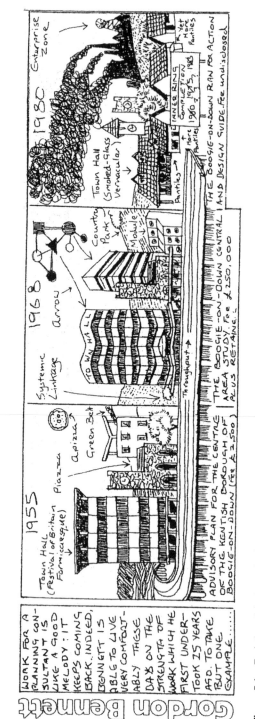

Figure 5.1 Evolution of design styles in post-war Britain. *Source: Planning, 409, 13 March 1981*

the evolution of the tower block from the 1950s to the 1960s and then the reaction of low-rise neo-vernacular postmodern buildings, which though traditional on the surface are in fact system-built inside. As Box 5.1 shows, these design elements were to some extent mirrored by changes in planning practice and theory.

Box 5.1   Mock exam questions typifying changing attitudes to planning

1958

Summarise the advice in Circular 58/51 and use it to advise whether conditions requiring proper construction of a crossing over the grass margin to provide vehicular access to a highway might be *ultra vires* if the grass margin is not in the applicant's control. (Emphasis on planning as physical design and control over minutiae.)

1968

Explain how conditions requiring construction of a crossing over a grass margin to provide vehicular access to a highway might be used as a cybernetic control mechanism satisfying the law of requisite variety, and steering the system away from negative entropy. (Emphasis on planning as a system of related elements.)

1978

'The state is merely the executive committee of the whole bourgeoisie' (Marx). Discuss this statement by referring to the class contradictions posed by a planning condition requiring a crossing to be constructed over a grass margin to provide private vehicle access to collective consumption goods such as a highway. (Emphasis on Marxism but only by theoreticians, not by practitioners.)

1988

A major business park development is stalled by a planning condition blocking vehicular access to the highway. On behalf of your client write a letter to the planning authority examining the impact this is having on job creation and suggesting how a Simplified Planning Zone might be designated to expedite development. (Emphasis on market-led planning and the use of planning consultancy firms.)

1998

A male motorist wants to tarmac over a valuable ecological corridor, so that he can further undermine the sustainability of the planet by driving. What do you think about the ethics of this development? Form yourself into groups and role-play to understand how the flora and fauna might feel. (Emphasis on sustainability albeit cosmetic in the face of economic imperatives but gender-driven humanism not reflected in practice.)

Source: Hague (1998). Authors comment's in parentheses at the end of each question.

Because British public policy tends to be incremental a common explanatory theme is to take an evolutionary approach in which each change is explained by defects in the previous arrangements caused either by implicit faults or by changes in the external environment. Since planning is strongly influenced by the economy one obvious approach has been to study how planning policies have reacted to economic cycles. For example, Hobbs (1992) in a detailed analysis of the post-war period emphasises the priority given to production by the system. He finds that planning policies responded to changes in the economy, so that they became weaker in times of recession and stronger in boom times. Ironically, in times of recession planning emphasised how it could improve environments at the local level. Thus Hobbs concludes that the system does give discretion to reflect uneven economic performance at the local level.

An excellent evaluation of the system till the late 1980s based on extensive research between 1978 and 1984 has been provided by Healey *et al.* (1988). This work concludes that the planning system has been universally accepted because there is both a need to assist the spatial co-ordination of individual activities and a need to mediate conflicts of interest over how land should be used. The outcome of mediation is the product of the way participants assert and confront the power relations reflected in:

1   The way in which specific demands of different groups for action from the planning system are constituted.
2   The instruments and institutional arrangements of the system.
3   The criteria which participants in the policy processes draw upon in determining action.

Although Healey *et al.* identified powerful groupings in the mediation process they stressed that these groups do not always assert their interests and that different coalitions of groupings are formed for individual issues. Nonetheless, a major analytical problem is that the powerful interests that are favoured are hidden within the complex and often routine assumptions of planning practice. Thus the power relations and ideology of previous periods lurk within the processes of the present, invisibly blocking the entry of new interest claims.

However, power relations do eventually change, and did so notably during the period of Thatcherism from 1979 to 1990. For example, in his book on Thatcherism and planning Thornley (1993) argues that planning was reoriented in three ways: first, greater acceptance of market criteria; second, the selective application of environmental criteria; and third, the removal of social criteria. There was also a reorientation of procedures away from local democracy towards centralised government supervision. Planning also became divided into three areas: first, planning controls remained strong in good environmental areas unaffected by the 1981 GPDO change (some 25% of England); second, in most areas economic criteria dominate; and third, in some areas, e.g. Enterprise Zones and Simplified Planning Zones, planning has effectively been bypassed. However, these changes were not an attack on the overall principle of managing land use change but were a critique of the policies and practices being used – a view that had already been put forward by Healey *et al.* (1988) in their analysis, above.

Allmendinger and Tewdwr-Jones (2000b) have also examined the Thatcher period but have extended their analysis into the epochs of John Major and Tony Blair. They contend that Thatcherism had three main tenets: the rule of law, centralisation and market orientation. Thatcherism thus contained one central contradiction, belief in personal freedom but also in strong central control from the centre when necessary. These three tenets were continued into the 1990s by the dominance of the New Right philosophy, as shown in Table 5.1. However, the importance of the market was played down in the 1990s, as the environment became a more important political consideration via international conferences like Rio, and voter concerns at home. The advent of Tony Blair in 1997 led to a further modification of the New Right and greater emphasis on institutional systems and performance-related targets in which planning had to display its cost-effectiveness to society, as shown in Table 5.2.

They conclude that New Labour is a pick-and-mix approach and that incremental change will be the order of the day. Within this theme there have been two main changes so far under Tony Blair. First, 'predict and provide' has been replaced with 'plan, manage and monitor',

Table 5.1 Changing attitudes towards planning under the New Right

	Approach	Means (in order)	Ends	Examples
1979–89	Project-led, reduce strategic role of planning and make decisions on their merits. Development-led	1 Market orientation 2 Rule of law 3 Centralisation	Deregulation of controls (release pent-up demand through targeting supply-side constraints. Make system more 'transparent' by reducing local planning authorities' discretion. Do not alienate conservation-minded voters	Urban Development Corporations. Enterprise Zones. Simplified Planning Zones. Circular 22/80, *Lifting the Burden*
1989–97	*'Plan-led'.* Reintroduce a (modified) element of strategic role for planning that takes strong central guidance into account. (Illusion of) locally led planning to reduce responsibility of Secretary of State	4 Centralisation 5 Rule of law 6 Market orientation	Control planning through strengthened central mechanisms. (Grudgingly) introduce environmental concerns and design criteria	Elevation of importance of Development Plans. Revised PPG series. Rio commitments. Section 54A, 'Our Common Inheritance'

Source: Allmendinger and Tewdwr-Jones (2000b)

and second, the 'sequential approach' has been introduced, in which applicants have to demonstrate a sequential search outwards from the city centre before the release of greenfield sites can be contemplated. In a further analysis Allmendinger *et al.* 2004) found four key changes since the advent of Blair in 1997:

Table 5.2 New Right and New Labour comparisons

New Right	New Labour
Rule of law. Remove/minimise discretion to allow greater certainty for the market to operate.	*Enabling and enforcing role for the state.* A 'hard edged' provider and enabler of rights and responsibilities built upon social justice and inclusion.
Centralisation. Suspicion of democracy and belief in authoritarian style.	*Individual and community focus.* Decentralisation of power and decision making to help enable greater community responsibility and functionality.
Market orientation. Market as the means and the ends	*Greater supply-side emphasis within strict monetarist approach.* Market as part of means towards individual freedom paying for greater social inclusion.

Source: Allmendinger and Tewdwr-Jones (2000b)

1   The orientation of planning towards community concerns at the local level.
2   The translation of land use planning to spatial planning.
3   A focus on planning as a co-ordinating or choreographing enterprise.
4   A determination to enhance participation within planning policy development at all scales.

In an excellent survey of the 1980s and 1990s Tewdwr-Jones (1999) argues that during the 1980s free-market appeal-led planning came to the fore. In the 1990s there was a move to a plan-led system driven by the top-down approach of RPGs with central government and big business handing down policies for rubber-stamping by LPAs. The result has been three types of planning. First, plan-led certainty in areas where development and the environment can be balanced. Second, flexible project-based policies in areas of urban regeneration, and third, regional economic and political planning achieved via collaborative planning among the 'great and the good' to promote regional growth against the wishes of local communities.

Table 5.3   The changing orthodoxy of planning

The previous orthodoxy	The new orthodoxy
Attitude to change and places	
Green Belt	Sustainable containment
Overspill/New towns	Transport corridor development
Countryside as agricultural production	Countryside as recreation, consumption and diversified economy
Provide space/reduce densities	Compact city/increased density
Growth	Sustainability
New build	Renewal
Betterment levy	Planning gain/impact fees
Place making	
Segregation of land uses	Mixed uses
Catering for the car	Restraining car use
Hierarchy of retailing	Vital/viable town centres
Land use standards for open space	*Open space as ecological resource*
Presumption of set standards	Recognition of special needs
Eliminate non-conforming business uses	Promote/foster small business
Redevelopment	Regeneration
Twelve-hour city	Twenty-four-hour city
Processes	
Plan with clear aims	Plan providing strategic framework
Map-based system	Policy-based system
Planner as technician	Planner as manager/co-ordinator
Separate planning departments	Planning as part of corporate unit
Public investment	Private partnership
Gender/race/disability-blind	Aware of equality dimension

Source: adapted from Roberts (2002)

In an overview of his work Tewdwr-Jones (2002) comes to two main conclusions about the changes in planning under the Thatcher, Major and Blair epochs. First, there is clearly a gap between central government expecting its national planning policies to be followed and local government merely taking central government's policies into account when making decisions. In other words, though national control is apparently strong via legislation and overt policy guidance via PPGs and appeal decisions, in reality local planners have considerable discretion in how they implement both national and their own local policies. Second, the political ideology from Thatcher onwards has been one of market orientation and centralisation. However, recent changes towards regional assemblies and the considerable powers forecast to be given to regional planning in the 2004 Act (see later in the chapter) indicate a clear departure from market-oriented, project-led planning and a move towards a regionally sensitive and locally oriented plan-led system.

From the perspective of an experienced planning consultant and former president of the RTPI Roberts (2002) claims that a new planning orthodoxy emerged over 25 years, as shown in Table 5.3. The key features have been a change towards a mixed-use 24-hour environment based on a multitude of activities, a move away from technical issues, and a move towards greater use of partnerships and social inclusion.

Finally, Cherry (1996) has provided an excellent overview of the century and in particular the post-collectivist attack on planning in the 1980s and 1990s which according to Hayek branded ideas that we could shape the world around us as a 'fatal deceit'. Planning has been neutered and is tolerated as a harmless low-key activity, which may be marginally beneficial as a balancing ploy among contenders in environmental protection. It has become short-, not long-term; incremental, not strategic; surprise-free; highly regulatory; and over-systematised.

Evaluation by typology

Jordan (2002) has developed a typology for assessing the efficacy of a political system using the basic divisions shown in Table 5.4. Thus the typology produces four types of evaluation:

Table 5.4 A typology of planning problems, policy goals and policy outcomes

	Focus of analysis	
Orientation to problem	Policy output	Policy outcome
Orientation to policy goals	1	2
Orientation to policy problem	3	4

Source: after Jordan (2002)

1 Extent to which policy outputs reflect policy goals.
2 Extent to which outcomes achieve policy goals.
3 Extent to which policy outputs address policy problems.
4 Extent to which policy outcomes have solved policy problems.

The typology can be adapted for four types of evaluation in the context of planning. First, the appropriateness of the legislation and policies for the land use problems they were designed to address. Second, the efficacy of the planning organisations and processes set up by the legislation. Third, whether the planning policies developed by the system are appropriate for the issues they were designed to address. Fourth, the impacts of the policies on the problems they were designed to address and also significantly the unexpected side effects of the processes, for example dramatic house price inflation. This typology provides a variant of the Morrison and Pearce (2000) sequential checklist modified and used later in this chapter, in the hybrid approach. Like Morrison and Pearce, Jordan does not go on to evaluate the system using his typology.

An example of where the typology approach has actually been used to produce an evaluation has been provided by Brindley *et al.* (1988), as shown in Table 5.5 in their review of the Thatcher years. Eight years later Brindley *et al.* (1996) argued that postmodern pluralism underlay Thatcherism and had now replaced it. Within the typology, private management and public-sector investment planning had been abandoned, leaving responsive planning (trend + regulative) and partnership planning (leverage + popular). Table 5.6 shows the key characteristics of these four styles of planning.

Table 5.5 A typology of planning styles

Perceived nature of urban problems	Attitude of market processes	
	Market-critical	Market-led
Buoyant area	Regulative planning	Trend planning
Marginal area	Popular planning	Leverage planning
Derelict area	Public investment planning	Private management planning

Regulative planning refers to planning by controls over applications. *Trend planning* refers to the following trends: 'predict and provide'. *Popular planning* refers to following the popular will. *Leverage planning* refers to borrowing capital, from the American word for borrowing. *Public investment* planning refers to using public funds to boost infrastructure. *Private management planning* refers to private capital being used in areas where planning controls have largely been relaxed in favour of their own rules.

Source: Brindley *et al.* (1996)

Table 5.6 Characteristics of four planning styles identified by Brindley *et al.*

Planning style	Institutions	Politics	Conflicts and tensions	
			Limiting factors	Principal beneficiaries
Regulative	Local authorities	Technical–political	Strength of local market	Local land and property owners
Trend	Local authorities	Non-strategic gatekeeping	Retaining any control of market	Incoming developers
Popular	Community organisations	Imperfect pluralist	Community control of resources	Local lower-income groups
Leverage	Quasi-governmental agency	Corporatist	Potential for market revival	Incoming developers

Source: Brindley *et al.* (1996)

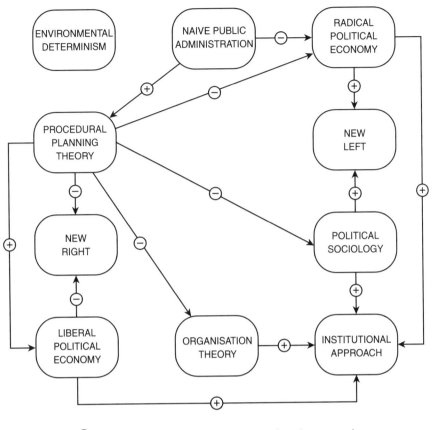

+ One approach developing positively out of another approach

− One approach developing as a critique of another approach

Figure 5.2 Relationship between theoretical approaches to planning. *Source:* Rydin (1993)

Rydin (1993) has also provided a useful typology based on the development of planning thought over time and shown how they relate to wider theoretical developments, as shown in Figure 5.2. As Chapter 2 has already shown, planning began with the view that better environments would lead to a better society. This 'environmental determinism' reached its zenith in the first 30 years of the twentieth century when the 'new towns' movement was at its height. Socialism was the 'zeitgeist' of the 1930s and 1940s, thus when planning really began in the 1940s it was based on a 'public administration' model in which the state was the prime mover and shaker.

By the 1960s systems theory and computers had taken men to the moon and so it was believed that systems theory could not only explain how cities worked but also predict their future. 'Procedural planning theory' was born, based on rationality and scientific prediction, and so the 1960s and 1970s were dominated by so-called rational planning. However, in the 1980s theory and practice went their separate ways. Practice moved towards 'liberal political economy', supported by a small group of New Right academics, and developed the ethos of pragmatic planning: 'getting things done' and 'making things happen'. Most academics, however, went down the Marxist road. The 1990s saw the revival of so-called 'plan-led' planning and rational planning was to some extent revived but under an 'institutional approach' dominated by professionals and judged by performance criteria and key indicators, as already shown by Allmendinger and Tewdwr-Jones (2000b).

Rydin uses these concepts to provide a matrix which involves four columns, the main views of economic processes and thus the role that planners should play, and three rows, the main actors in the process, as shown in Figure 5.3. In the 'ideal market' model planners act as referees. In the 'market failure' model planners act to improve the market. In the Marxist model planning acts to achieve social equity and in the 'institutional economics' model, in a partial throwback to the 1970s, planners act as corporate computer men. It is this model that is applied by New Labour in its emphasis on creating a 'New Britain' based on a rational review of its institutions, and as Figure 5.4 summarises it can be seen as the end result of related changes in the twentieth century.

Evaluation by concept and theory

In addition to the theoretical analyses employed in Chapters 2–4, concepts and theoretical approaches have been applied to the wider context in which planning finds itself, notably the changing views of society and its relation to power. A number of concepts have been used, notably sustainable development and ecological modernisation. Two main theories have been used, based on the ideas of Foucault and Habermas.

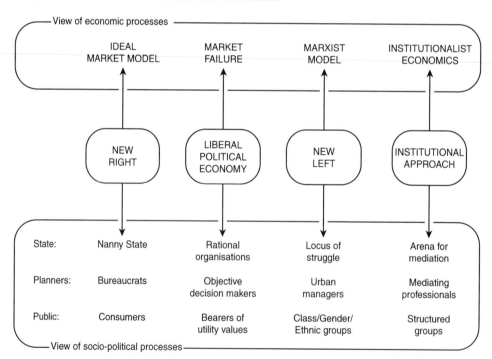

Figure 5.3 Theoretical approaches based on a view of the economic process. *Source*: Rydin (1993)

	Nineteenth and early twentieth centuries	1920s, 1930s and 1940s	1950s and 1960s	1970s	1980s	1990s
Economic and social change	Industrialisation Urbanisation War	Recession and restructuring War and reconstruction	Post-war boom Mixed economy Consensus politics	Turning point in economic growth Urban–rural shift Inner-city decline	Recession (and recovery) New technology Collapse of mixed-economy consensus	Globalisation of: politics, economics and environmental change
Salient political issues	Public health Social unrest	Regional unemployment Suburban growth	Increasing living standards Rapid development	Racism and urban disorder Excesses of economic growth	Unemployment Track record of public sector	European integration Environmental crisis
Key planning activities	Housing Public sanitation	Regional planning	New towns Redevelopment	Inner-city policy Rehabilitation and conservation Pollution control	Urban regeneration Countryside policy	Regeneration Sustainable development Flagship projects
Planning profession	Architects Engineers	Growth of separate identity	Corporate planners	Crisis of competence	Retrenchment Privatisation	Reassessment
Theoretical framework	Environment determinism	Emergent planning theory	Procedural planning theory	Critiques: Organisation theory Welfare economics Radical political economy Urban politics/sociology	Political ideologies New Right New Left	Collaborative Planning Critiques (reprise): Environmental economics Radical political ecology Environmental justice
Conceptualisation of planning	Urban design	Public sector direction of land use	Generic decision making	1. Policy implementation 2. State intervention 3. Community empowerment	1. Economic development 2. Community empowerment	1. Place making 2. State intervention 3. Community empowerment

Figure 5.4 The development of planning policy and theory. *Source:* Rydin (1998)

Concepts

The concepts of sustainable development have been examined by Healey and Shaw (1994), who in a textual analysis of the discourse contained in post-war planning documents have identified five strands in which the environment has been treated by planning:

1   A welfarist–utilitarianism approach combined with a moral landscape aesthetic (1940s and 1950s).
2   Growth management, servicing and containing growth while conserving open land 1960s).
3   Active environmental care and management (1970s to date).
4   A marketised utilitarianism, combined with the conservation of nationally important heritage (1980s to date).
5   From the 1990s onwards, the rhetoric of sustainable development.

These strands have contained three types of approach. First, viewing the environment as a functional resource, a reserve of non-renewable resources and amenities for human enjoyment, which Whatmore and Boucher (1993) have shown treats the environment as a commodity in the process of negotiating environmental 'planning gains'. Second, the moral and aesthetic notion of treating the environment as a backcloth or setting to human life. Third, the environment being seen as a constraining factor. Although the five strands and three approaches have coexisted with each other and other planning policies, it is the modernist rational discourse of economic growth, instrumental rationality, and the ability of the human species to control and manage nature, which has won out.

This is a view endorsed by Blowers (2000), who in a review of sustainability argues that the concepts of sustainable development have been appropriated by the interests that shape social relations in modern society. The rhetoric of this transformation includes notions of consensus, co-operation, collaboration and above all partnership. Blowers argues that two new paradigms have arisen as part of this transformation: political and ecological modernisation. Ecological modernisation has achieved dominance in political circles because, in line with the rhetoric outlined above, it suggests that the relations between the economy and the environment can be complementary rather than in conflict.

Political modernisation (PM) has four key features. First, an emphasis on the market. Second, a distinctly different form of governance, both centralised and devolved but with a decline in representative democracy, notably at the local level. Third, an emphasis on collaboration to be achieved by negotiation, and fourth, the reduction of state control by a 'hollowing out' of the state. Political modernisation is a system that is exclusive, elitist and unrepresentative. Within PM planning occupies a marginal but paradoxically pivotal position as a clearing house for pressures from environmental groups.

Ecological modernisation (EM) has three antecedents based on environmental policy making. First, a long-standing concern with conservation and preservation, second, a concern with public health, and third, concerns about global change. However, the market still rules, so the only serious question is how best to manage the environment for capital accumulation, economic efficiency and growth. Ecological modernisation is thus a mode of policy making which upholds the market while at the same time recognising the need for regulation to prevent environmental degradation. More positively it is seen as a potential source of future growth.

There are four basic components of EM. First, the necessity of incorporating environmental issues into production and consumption processes, for example recycling. Second, the prominence of the market. Third, the role of the state under EM is seen as enabling development by a framework of controls and incentives via regulation. Fourth, under EM environmental movements are increasingly involved in collaborative relationships in policy making with business and government in an 'iron triangle'.

In terms of planning, Blowers believes that EM has three limitations over and above being a prisoner of prevailing social and political conditions. First, planning lacks effective powers and can prevent but not promote development. Second, strategic planning has a weak institutional base, with growing conflict between RDAs and RPGs. Third, planning is limited in its reach to land use.

But according to Blowers planning can react to two issues. First, the failure of EM and PM to address social inequality, intergenerational problems and environmental deterioration. Second, it can work with counter-modernist movements sympathetic to the enduring social purpose of planning and the need to value nature more. Thus planning could be a vehicle for change with six main attributes. First, planning is concerned with the value of natural surroundings and the conservation of both the natural and the built environments. Second, planning links the local with the global by stressing land use issues. Third, with regard to intergenerational issues planning takes a long-term view. Fourth, in terms of empowerment planning has a primary role in promoting the public interest. Fifth, planning like sustainable development stresses the need for a holistic approach. Sixth, planning needs to reinvent its social purpose and its concern with inequality and social justice.

In reality the tendency has been to contain the planning agenda to a narrow remit focused on land use and development and to a discourse dominated by forms of utilitarian functionalism. The challenge for the new environmental agenda is to gain real leverage over economic discourse, not by the reinforcement of traditional strategies and policies but by a fundamental rethinking of the form and content of the system in terms of conceptions, technical methods and policy processes. In the meantime Healey and Shaw (1994) have concluded that planning is very much an administrative process:

The institutional arrangements and policy tools of the planning system appear, therefore, to have the flexibility to accommodate the task of regulating the use and development of land in such a way as to achieve the objectives of sustainable development. The key challenge of this task is to work out how demands for development and resource exploitation can be combined with the environmental objectives of handing on to future generations an environmental inheritance equivalent to that received by the present. This task has both material considerations – the balancing of one set of objectives against another – and moral dimensions in terms of value placed on environmental qualities and relations with the natural environment. Thus the planning system appears to provide a flexible regulatory regime: a set of formal mechanisms and practices built around them through which the tensions between the potentially conflicting objectives of economic development, meeting social demands and needs, and conserving and enhancing environmental quality may be managed. Its traditions are dominated by an administrative-legal discourse within which judgements combining knowledge and value are typically intertwined.

(p. 426)

Theories

The main debate has been between theories focused on the work of Habermas and Foucault, outlined in Chapter 2, and used by certain authors in Chapters 3 and 4. Mantysalo (2002) has provided a review of the debate and in particular a critique of the theories of planning derived from Habermas. Mantysalo traces the way in which key writers like Forester, Healey, Fischer, Sage and Innes have used the critical social theory of Habermas to produce 'communicative planning theory' or what Mantysalo prefers to call 'critical planning theory'. In this process, normative issues like legitimacy, inclusiveness and quality in argumentation have become central in planning theory debates. Critical planning theory thus addresses planning practices both descriptively and prescriptively – grasping the essence of planning as a problem-solving activity that transcends rationality and necessarily manages social relations. However, Habermas's conceptual separation of communicative and instrumental rationalities, and his total reliance on rationality, make such theoretical work inherently problematic. In order to add descriptive and prescriptive capacity, planning theorists have had to look for other theoretical sources, such as Foucauldian power analytics.

A key figure in the debate has been Patsy Healey. In particular she is concerned that planning should not be seen as merely regulation and has called for a more inclusive approach in which all social groups are included and people's needs and wants are given much more consideration in a system of so-called 'communicative' or 'collaborative' planning (Healey, 1992). There are two strands to the argument. First, that planning can best be understood by a discursive and qualitative approach, and second, that planning should embrace an approach that engages with people. Healey spelt out the key themes in her 1997 book *Collaborative Planning*, which traces theories of planning from rational process models (1960s–70s) and Marxism (1970s–80s) to institutional (Giddens) communicative (Habermas) models of the 1990s. These theories have not, however, been reflected in the way in which planning has been operated on the ground, as shown diagrammatically in Figure 5.5. This shows that the latter half of the twentieth century

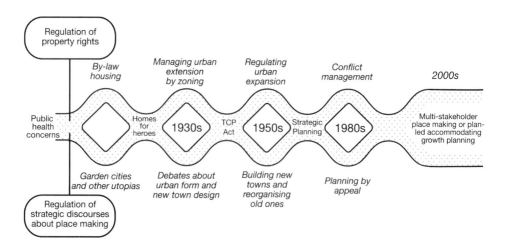

Figure 5.5 The evolution of regulatory purposes in the twentieth century. *Source*: adapted from Healey (1998)

was dominated first by regulation and then by conflict management from the 1980s onwards, with the possible emergence of a new paradigm of a collaborative, multi-stakeholder place-making mode for the new century. Under this mode, planning should thus become based on a series of ongoing communications between all actors involved.

In a further development of her vision Healey (1998) has linked collaborative planning with 'making places'. This strategy is an alternative to the more mechanistic approach of conflict management based on regulation, reconciliation and norms. In contrast 'making places' is all about discourse, flexibility and social inclusion in a view of planning based on people's needs and wants rather than the needs of the economy. But in a traditional search for consensus Healey also demonstrates how 'effective spatial planning' may lie at the centre of four types of planning, as shown in Figure 5.6, in spite of the differences between conflict management and place making shown in Box 5.2. The route to or from the centre of Figure 5.6 may, however, be twofold. First, depending on the circumstances planners may start from the centre and then pick the most appropriate of the four types of planning. Second, planners may use two or more of the four types in a search for consensus based on the creative tension between them.

Box 5.2   Differences between conflict management and place making

Conflict management	Place making
(performance criteria)	(collaborative planning)

Regulatory object

Right to use and develop land	The way of thinking; policy discourses; frames of reference

Context

Efficient land and property development markets; conflicts between development and environmental interests	Multiple stakeholders; complex political claims for attention; markets prone to failure

Discourses

Norms, standards and criteria; political legitimacy and legal interpretation; public interest versus private rights	Qualities of places; spatial organising ideas; strategic projects; coordination and collaboration

Function of plans

Product: store of norms, policy statements, policy criteria and so forth	*Product:* store of 'framing' principles about quality of places; and supporting arguments
Process: arena for conflict mediation	*Process:* collective social construction of strategic organising ideas, stories, images and so forth

Source: Healey (1998)

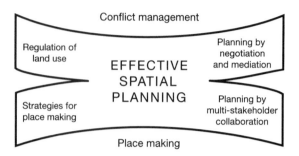

Figure 5.6 Conflict management and place making as alternative strategies for planning. *Source*: adapted from Healey (1998)

Healey's work has been welcomed as a renewal of the utopian vision of planning but has been criticised for being too idealistic and impractical, notably by Tewdwr-Jones and Allmendinger (1998). In response Healey (1999) ironically stressed the power of institutional procedures when she argued that, judged by Habermasian discourse ethics or evaluated in terms of Foucauldian power analytics, land use planning practices in Britain have become deeply embedded in routines of state procedure and allowed to privilege certain interests over others. In contrast, communicative practices and consensus-building strategies are rapidly expanding in the fields of environmental conflict resolution. For the analyst of governance processes Healey then raises the question as to why the practices of the system, which engages with multiple local environmental conflicts and involves a wide range of stakeholders, have not developed more communicative pathways in Britain. This suggests that there are some peculiar characteristics of the way the British land use planning system has evolved in the 1980s and 1990s which have constrained its trajectory (Healey, 1998; Vigar *et al.*, 2000). Faced with increasing local government conflict, the system has become configured as an introverted regulatory function of the state, with discourses and procedures honed for conflict resolution in semi-judicial contexts.

Healey advocates the greater use of consensus-building strategies in environmental conflict resolution as a better model. Vigar and Healey (2002) have also stressed the need for key strategies if the environment is to become a greater feature in planning policies. These strategies should be based on five key principles: effective articulation of the issues; framing the issues by stakeholders; co-ordination of stakeholders; ensuring that the environment is a legitimate policy aim; mobilising stakeholders and others into action.

For the future Madanipour *et al.* (2001) argue that dramatic changes in political and economic contexts have brought forward the need for spatial planning systems and practices to go through some substantial changes. In particular, there need to be new ways of 'doing governance' linked with new ways of thinking about space, place and territory if spatial planning systems are to be transformed into valuable governance activities in the twenty-first century.

Taylor (1998) has provided a complete overview of these theoretical themes and has come to the important conclusion that planning is neither science nor social science, but a form of social action directed at shaping the physical environment. Planning is thus driven by moral, political and aesthetic values and so planning is an ethical practice. This is, however, a fairly recent view. As long ago as the 1950s planning was about urban design before it was transformed by the scientific paradigm of the 1960s into a rational decision-making model based on systems theory. In the 1970s and 1980s planners came to be regarded less as technical experts than as a kind of facilitator drawing in other people's views and skills before attempting to achieve a consensus.

By the 1990s Taylor argued that there were two related views about the nature of planning, both rooted in the central concept that town planning exists to improve the world. First, town planning is a form of social action or social practice based more than anything on sound judgement. Second, judgements about what best to do are value judgements about the kind of environments we want to protect or create and therefore planners should undertake rigorous research into what is regarded as environmental quality in their areas.

In conclusion there is a major division between those who are happy to see planning as a regulatory process carried out within an actor-network framework and those who would like to see a more vibrant and inclusive process.

A HYBRID EVALUATION

Chapter 2 and the preface developed the view that the best way to evaluate planning is to take a hybrid approach based on arguing from the general to the particular and then back to the general from the particular, using the analogy of an egg-timer. Accordingly, Box 2.4 was produced using the framework outlined at the outset of Chapter 2 to organise the key evaluation issues that had been discussed in Chapter 2. Now that the particular issues in planning have been discussed in Chapters 3–4 and specific evaluations have been discussed in this chapter it is time to revisit Box 2.4. This evaluation rests on two basic premises. First, a review of published literature on planning and the use of an evaluation framework has allowed a set of key evaluation questions to be produced. Second, writing Chapters 3 and 4 and the first part of Chapter 5 has given me the insight added to 35 years of studying planning to make qualitative judgements. These judgements cannot be scientifically tested, and you the reader may well like to disagree with some or most of my judgements. But planning is a complex set of arguments and at the end of the day can be evaluated only like a court case. I will be the judge and you will be the jury. Up till now you have been presented with the evidence. Now I will present my summary. The verdict is of course yours.

In order to help an understanding of the next few pages it is suggested that Box 2.4 should be photocopied and kept by the side of the book.

1 Planning and society

To what extent is planning a mirror of its society? Yes and no. 'Yes' in that planning is part of mainstream politics, but 'no' because planning is a middle-class activity favouring the development of private property and protecting middle-class countryside areas. This is unlikely to change, because society has favoured the private sector since the 1980s.

Is the key role of planning to manage change with the minimum of conflict? Planning is tolerated as a service activity that reduces tensions within the pretence of democracy. The basic principle of economic growth cannot be challenged, so planning can only claim to manage change, where economic pressure is high, more efficiently than the private sector. In areas with little economic pressure it may 'make things happen' by co-ordinating land assembly and by relaxing controls.

Whose value systems decided which issues were addressed? Originally the value systems were those of planned development, mainly in the public sector. But this was gradually abandoned in the 1950s and the crossover to private development took place in the 1960s. Since the 1980s the values of the free market have been in the ascendant. However, since 1990 residents in attractive and prosperous areas have hypocritically espoused conservation values in order to protect their property values and environments. The clash between the demand for growth and conservation in these areas, notably south-east England, will present planning with its biggest challenge. Overall, though, the need for development has always been paramount, given politicians' belief that elections are won or lost on the state of the economy. In spite of a massive rise in living standards politicians still believe people want more material possessions. Thus it has not really mattered whether Labour or Tories have been in power, since both have demanded that planning must give precedence to economic growth.

Democratic versus autocratic. Planning decisions are nominally democratic and at the local level some decisions are the subject of fierce and genuine debate. However, major decisions are made by cabals of the 'great and the good' behind closed doors before the pretence of public debate ensues.

Public versus private. The public sector has declined drastically since its heyday in the 1940s and 1950s. However, some vestiges of public planning survive, notably in the concept of land ownership or management in the protected areas and land assembly in the most deprived inner-city or mineral waste areas.

Lay versus professional. At the local level planning decisions are made by planning committees of lay councillors, but most of the key decisions are effectively made by planning officers, albeit in consultation with councillors. In certain circumstances – notably major new village developments, for example – lay people will group together to form local protest groups, but they are rarely successful, since the key decisions have already been made further up the plan hierarchy.

Intellect versus values. Planning is a complex and abstract activity, especially at the plan-making level. It thus demands an intellectual approach. However, because it is so complex, value judgements are often intuitively made and then *post hoc* rationalised by intellectual arguments.

Ecology versus economy. This issue is repeated below and is indeed one of the key issues in planning. Some planners have tried to defuse the debate by invoking 'ecological modernisation' under which economic growth can take place without compromising ecology by increasing efficiency. However, this ignores the key fact that development takes up land and thus reduces the area of land for ecology. In addition, in 2003 the Airport White Paper accepted a threefold forecast in air travel with enormous increases in air pollution and implications for land use planning, for example more houses and road traffic, with little reference to ecology.

Local versus regional. Most people perceive planning issues only at the local level when a proposal impacts on them. However, the region is a natural area for planning, given a road-based society where goods are delivered nationally and people commute long distances. There is thus a major need to get people to engage with regional issues.

2 Planning objectives

Balance and sustainability

To balance protection of the natural and built environments with the pressures of economic and social change. The key issues have already been discussed above. The proposition is politically popular, since it implies that we can have our cake and eat it, but it is too vague in that it does not explain how balance may be achieved. It is wonderful rhetoric and reinforces the belief that the planning system relies heavily on judgemental decision making, since no mathematics can provide the necessary balancing equation. What really happens is that some wildlife sites are preserved and some buildings maintained as long as the economic activity involved is allowed to happen. The balance is nearly always unequal in favour of economic growth. However, in 2004 Dibden Bay, a major new port on Southampton Water, was turned down on ecological grounds after a public inquiry.

Twin pillars of managing urban growth while protecting landscape and heritage. The post-war planning system was based on an alliance between three groups: house builders, farmers and conservationists. House builders were allowed to build but only in designated areas, farmers were encouraged to produce in the knowledge that their land would not be taken, and conservationists were satisfied that most of the countryside would be protected. This fortuitous alliance has been tested since the 1980s as the need to protect farmland has diminished. The conservation lobby has now forced house building on to the defensive and house building has been restricted to its lowest rate for a century. This, however, is largely the result of extreme local pressure against development and does not undermine the hegemony of economic growth outlined above. Indeed, the government is taking steps to overrule local democracy and impose major house building in several areas.

Ensure sustainable land use. Sustainability has become the sacred cow of our times and has largely subsumed the two objectives discussed above. However, sustainability has also become a word that means everything and thus nothing. For example, the plan to impose major new house building in the South East outlined in Chapter 4 is called the 'Sustainable Communities' programme, but this is a gross misuse of sustainability. The new settlements will be based on suburban environments where people will have to travel by car for all their needs and wants. In contrast, genuine sustainable developments based on the use of local materials and resources are consistently turned down by the planning system. Sustainability is an excellent objective but it is being sadly abused by its present application.

To set a land use framework for economic development

To ensure provision of land for jobs. To provide housing for all sections of society. To provide safe and efficient transport systems. To ensure the best use of mineral resources. Reduce deprivation and encourage regeneration. Maintaining town centres and revitalising older urban areas. This objective is the one that matters because it concentrates on the role of planning in helping economic growth. In order to make it more palatable, social objectives are tacked on but the real message is that planning is about facilitating, not preventing or imposing conditions.

To fulfil the vision of development plans

Co-ordinate investment in land and buildings. To make things happen at the right time and the right place. To ensure development works in a functional sense. Planning can help society by managing change more efficiently, in particular by co-ordinating jobs, housing and transport. It is an attractive objective but one that it is very difficult to achieve, given existing infrastructures and that the impetus for change comes from the private sector. Nonetheless, it is inconceivable that the attempt should not be made, since the alternative free-market scramble is too awful to contemplate.

Improve quality of life

Protect the architectural and cultural heritage. To ensure development is visually pleasing. Sustaining the character and diversity of the countryside. Defending Green Belts to prevent sprawl. Improving the quality of residential development. These objectives are very appealing and allow planning to be sold to the general public as a 'good thing'. However, in reality most people interpret it as a service for protecting neighbourhood interests and resolving disputes with their neighbours over mundane, albeit important, issues like high hedges and extensions blocking light or invading privacy.

3 Outputs (plan making)

Proactive reconciliation. A perfect plan would predict land use conflicts and reconcile them at the plan-making stage. Attempts to do this have been made in Switzerland. In reality, the future is too uncertain and people are too apathetic until development is nigh to concern themselves enough with hypothetical scenarios. Nonetheless, it seems eminently reasonable that the attempt should be made by consensus seeking during plan preparation.

To what extent should planning evaluate itself through its cyclical process of plan preparation? A crucial element of plan making is its iterative and cyclical nature, but to a large extent plan makers are judge and jury in their own domain. Most plans are subject to scrutiny in a public inquiry but these are incestuous affairs. There is thus a case for an external body to assess not just plan making but also development control on a number of criteria by sampling plans and development control decisions. This would be in line with our 'audit'-driven culture but might undermine professional self-confidence.

Are land use plans the most efficient way of promoting land use policies? Originally the plans were plans for the public sector and were thus effective blueprints. Since the 1960s plans have largely had to rely on the private sector to develop in the selected locations and although the private sector is consulted during plan preparation they may not get the location they desire. Plans are also too static and often out of date before they are put into operation. However, plans do give clear guidance and as legal documents they give planning legitimacy. Plans are not perfect vehicles but they are the best we have.

Should decision making on plans be more democratic and be transferred from experts? Plans are not subject to much democratic decision making and are debated only at three stages. First, the general setting of overall targets, second, the draft plan, and third, the final plan. Most of the work is devolved to officers working with expert committees of councillors. Given the apathy of the public over abstract concepts like plans it is hard to see how plan making could be more democratic. Public participation exercises are expensive, long-winded and dominated by middle-class groups. Plans suffer from a democratic deficit but no obvious alternative exists.

4 Outputs (development control)

Reactive reconciliation or managing change by bargaining/consensus seeking? Planners can act like a judge or a referee. As a judge they can seek reactive reconciliation as in a civil case by attempting to resolve the issues by imposing conditions on a proposed development in order to make it more acceptable to opponents. As a referee they can involve themselves in the game by encouraging the players to interact all the time, not just when a decision is needed. In this role planners become an integral part of the development process, seeking to build coalitions and networks between the stakeholders. Although more time-consuming, and potentially open to charges of corruption or favouritism, this proactive role should offer a genuine opportunity to provide environments that are favourable to the widest community.

To what extent should development control decisions be based on land use policies? The hallmark of the British planning system is the discretion it gives decision makers. This reflects the pragmatic view of public policy espoused in Britain. However, it reduces the certainty desired by developers and conservationists alike that a zoning system would give. The Dobry solution offered in the 1970s could well be a way forward. Simple, local applications and those that conform to plans for the area could automatically be given permission if no substantive objections were made within say two months. All other applications would be subject to existing procedures.

Should development control decisions be more democratic or transferred to expert committees? At the moment most development control decisions are made by officers and accepted on the nod by monthly committee meetings. The selection of which applications get to be debated is somewhat random and often dependent on pressure from individual councillors or pressure groups. As in the previous section, applications could be divided into two streams, ones for democratic debate and ones for expert committees. The current system does not allow for third-party appeal, on the grounds that this would introduce lengthy delaying tactics. This does not seem to be democratic, and so a system based on a minimum number of appellants objecting to a planning permission could be set up.

Is the twin system of plan making and development control an efficient delivery system? In theory the two systems are inextricably linked and axiomatically development control needs a set of policies in order to give the coherence and certainty that planning claims to offer society. In practice, plans are either out of date, do not even exist, do not contain policies relevant to the application or, more often, contain too many conflicting policies. There are two alternatives. First, make plans much simpler with an annual review. Second, delay development control decisions until the plan-making system can produce relevant up-to-date policies. Neither of these seems realistic, so the best solution would seem to be the *status quo,* albeit with reform of the plan-making system to gear it more closely to the needs of development control.

5 Outcomes

Were policy responses pragmatic and incremental reactions or ideological and radical? The responses were very pragmatic and incremental, although the Thatcher period saw a number of failed attempts to institute radical reform. However, the past 20 years have seen economic issues become even more important in spite of an increased rhetoric of sustainable development. Planning decisions have very much reflected the spirit of the time in which they were made and the location that they were made in.

Has the system been effective as a system and also cost-effective for society? Plan making has been inefficient and ineffective because it has been too slow and contains too many policies. Development control has done a marvellous job in patching up the system with its emphasis on treating 'every case on its merits'. Planning is cheap to run and has been a very cost-effective benefit for society, preventing many bad things, making poor things better, but not often producing high-quality environments. Although development control is repeatedly accused of delay most decisions are made within two months. Regional and unitary city authorities should aid plan-making efficiency, albeit with a reduction in democracy.

6 Impacts

Is evaluation possible? Not really, given the complexity of environmental change and the difficulty of accounting for the 'birth control' and 'displacement' effects. However, some attempts have been made and as long as each evaluation is taken with the 'health warning' that the evaluation is only as good as the evidence then evaluation is possible. Evaluations of the process are more possible than evaluations of outputs and impacts.

Has enough research been done on both the system and its impacts? Far too little, and apart from Peter Hall's massive study in the 1970s there has been no major evaluation of the impact of the system. Some research has been done on detailed aspects of the system, but too much of it has been too theoretically driven or over-concerned with day-to-day practices.

Impacts based on objectives for planning set out in item 2 above

These are broadly written under the same headings as objectives but are modified to focus on whether the objectives have been achieved rather then the legitimacy and practicality of the objectives as discussed in the earlier section.

To balance protection of the natural and built environments with the pressures of economic and social change. The pressures of economic change have proved to be irresistible and the scales have been heavily weighted in favour of economic development, although housing starts have been severely cut back since the early 1990s. Social change has been reflected in a massive move towards free-market suburban housing which from repeated surveys is said to reflect consumer demand, although the nuclear family best suited to such housing is now no longer the majority household. New types of urban design to reflect the growing number of single households have not been developed by the system. The jewels of the natural and built environments have by and large been protected but ordinary environments suffer from neglect or the intrusion of banal standardised buildings.

Twin pillars of managing urban growth while protecting the landscape and heritage. Urban growth has been contained into discrete areas and there has been little sprawl or sporadic building in the open countryside. *This is the one great triumph of the system.* Important landscapes and heritage areas have been well protected but not enhanced. Ordinary landscapes have suffered from the lack of controls over farming and forestry which have damaged traditional landscapes through the clearance of hedgerows and orchards and the planting of over a million hectares of mainly alien species of one age and one species conifers.

Ensure sustainable land use. This has perhaps been the biggest failure of the system. The planning system has allowed, perhaps even encouraged, the spatial division of work, home and play, against the central principles of the self-contained community espoused by the new towns movement. This has created a road transport-dominated society and economy in parallel with the United States but fortunately not to the same excessive level. Nonetheless, our day-to-day lives are dangerously dependent on fossil fuels and, as disruptions of energy supplies have shown, everyday life breaks down quickly if energy supplies are cut off for even a few hours or days. People are now accustomed to travel long distances for work, shopping and play. This is partly due to cheap fuel and partly due to the division of space into one-type land use zones as advocated in most land use policies. It is sustainable only if cheap energy remains available and if global warming does not catch up with us. Planning may not be able to prevent these trends, however, since they are also due to globalisation and social change. By restricting car parking spaces and encouraging development around public transport foci planners can encourage more sustainable lifestyles.

To set a land-use framework for economic development

To ensure the provision of land for jobs. To provide housing for all sections of society. To provide safe and efficient transport systems. To ensure the best use of mineral resources. Reduce deprivation and encourage regeneration. Maintaining town centres and revitalising older urban areas. Although right-wing economists claim that planning is a drag on economic growth, largely in delaying development, most studies have found that businesses do not regard planning as a major constraint, but as one factor to be considered. Planning has thus provided a positive framework. Jobs have been provided even in protected areas and the nation is well housed. However, growing constraints on house building have contributed strongly to the dramatic rise in house prices over the post-war period. Land use planners do not control transport systems, although there have been unsuccessful attempts at the national level to integrate transport and land use planning. Land use plans are, though, dominated by consideration of transport systems, and many plans are strongly influenced by transport patterns and nodes. The failure of roads to keep up with traffic growth has led to massive congestion. This can be partly blamed on planning decisions, which have separated land uses, notably suburbia, thus necessitating car travel and making bus travel unrealistic.

Mineral resources are better managed than they used to be, notably with regard to restoration after exploitation. Edge-of-town tower blocks dating from the 1960s continue to be a major problem, but since they were built by housing departments which did not need planning permission from their own councils they are not really the fault of the planners. Town centres have been maintained in a judicious balance with of out-of-town retailing. The fate of deprived inner-city areas is a regional lottery. Where gentrification is desirable houses have been renovated but in large parts of northern industrial cities houses have become virtually unsellable. These processes are largely outside the power of the planning system. Planners cannot sensibly refuse renovation and they have no power to impose controls over those who live in the houses, although they can impose the condition that some are made available at social prices. Likewise, planners cannot force people to buy houses in derelict areas of inner Newcastle. In conclusion, planners have provided a framework for those people and those areas where economic growth is desirable. They can set the stage, but private-sector actors have to make the play happen. In areas where conditions are not favourable the stage can be set, but actors may not appear. Planning cannot be blamed as the main reason for the failure to regenerate deprived areas.

In conclusion, planning provides a permissive framework by identifying areas of growth. By and large, growth has taken place at such a pace that transport systems cannot keep pace. This is not a failure of land use planning but a failure of concepts of integrated planning across all sectors of the economy and a failure to invest enough in transport systems. If planning has failed in this area it is due to the separation of land use activities and the consequent need to travel from one zone to another.

To fulfil the vision of development plans

Co-ordinate investment in land and buildings. To make things happen at the right time and the right place. To ensure development works in a functional sense. The transport example above demonstrates that planning has perhaps not succeeded in co-ordinating development, but to expect it to is to overestimate what planning can do. To reiterate, planning can only provide a stage and guide and shape the movements of actors upon the stage. If the actors do not turn up there is little the planners can do. If too many actors turn up and too many directors try to manage the play the system becomes unmanageable. In the overheated south-east of England this is what has happened and national decisions over, for example, airport expansion have produced the sort of growth that is difficult to manage. The main problem is that modern life is too complex and too rapidly changing for the planning system to comprehend it, let alone manage it, given its limited resources and powers. Given these problems, the planning system has done an impossible job as well as possible and the economy continues not only to function but to grow.

Thus planning has acceded to globalisation forces and local economies have been under threat for a long time. The soulless rabbit hutches on postage stamps scattered round the edges of our towns and cities are a terrible testimony to short-termism and our frightening reliance on cheap road travel, electricity and telecommunications. Maybe the biggest challenge facing mankind is to reinvent local economies, because the whole world cannot live like Europe and the United States, since they depend on the resources of several planets to sustain their lifestyles. This will impose incredible strains that planning may help to diffuse enough to make the transition to a sustainable and fair world economy.

Improve the quality of life

Protect the architectural and cultural heritage. To ensure development is visually pleasing. Sustaining the character and diversity of the countryside. Defending Green Belts to prevent sprawl. Improving the quality of residential development. The quality of life varies dramatically between different parts of Britain. The architectural and cultural heritage of cities like Bath and Edinburgh has been protected, albeit with one or two notable intrusions. Development is not often visually appealing. Workplaces are too often banal metal boxes on soulless estates, although city-centre offices have moved away from glass tower blocks to sometimes visually stunning creations providing striking punctuation to the cityscape. Planners have had few powers to stop the destruction of countryside features but they have been guilty of allowing standardised designs into our rich diversity of regional architecture, even though imposing traditional designs may increase costs and have social implications. Green Belts have been preserved, but the design of suburbia remains dismally low, with unimaginative house designs sitting on road-dominated layouts. In conclusion, planning has become very heritage-friendly after initially allowing as much damage as Hitler's bombers. Indeed, planning may now have become too conservation-minded and the neo-vernacular pastiche nature of the suburbs is not to everybody's taste.

Wider issues

To what extent can the outcomes and impacts of planning decisions be disentangled from other decisions in society to isolate the effect of planning on environmental and social change? The transport and renovation issues discussed above illustrate the difficulty of disentangling the complex factors at work in environmental change. It is important to restate the fact that land use planning is one part of a complex system with very limited positive powers. However, the power to refuse permission or to impose conditions makes planning an important gatekeeper in the process of environmental change. The most important impact since the early 1990s has been the aggregate decision to restrain the number of new houses being built. This has contributed to serious house price inflation, but other factors like the growing number of households due to social change has also been at work, as has been the intense media interest in house price inflation. Planning is clearly a factor in change but not often the key factor.

What have been the main impacts of the system, both intended and unintended? The main intended impacts have been the containment of urban areas and the protection of the countryside from sporadic development. The main unintended impacts have been inflationary house prices, the separation of work, home and play and the resulting long journeys and congestion.

What were the main issues that should have been addressed by the system? With hindsight postwar affluence and notably the dominance of the motor car and leisure travel should have been foreseen. Thus an environment based on mixed land uses should have been the main aim of planning, rather than the emphasis on separating land uses into single activities. This would have reduced the need to travel, enhanced community spirit and created more diverse landscapes. It has taken us too long to realise that employment sites are no longer incompatible with residential uses.

Is the British system of democracy, based on incrementalism and accretionary precedent, a recipe for disaster for modernist planning? Planning is a modernist concept in that it attempts to create order from chaos by observing, explaining and predicting behaviour. However, the modern world is complex and restless and there is not enough time to understand one evolutionary phase before another one takes over. Plans that take years to prepare are thus out of date before they are approved. The overall system is changed only every 15–20 years, given the pressures on parliamentary time, and the changes can take over 10 years to implement at the local level. This may not, however, be a bad thing, since plans should not be breathless pop songs warbling to the latest fad, but should be taking a long-term view. The basic decisions about where land should be developed and where land should be protected should be based on longterm values of the land involved and the existing infrastructure. Thus muddling through in the short-term within the context of long-term goals may be no bad thing and may produce the eclectic landscape we have inherited from the past. For instance, English villages are so attractive because they contain a variety of designs and styles. Our goal now should be to create diverse townscapes rather than the monotonous banality of current construction.

7 Planning and society: four objectives, three big questions and the future

We begin with the four basic objectives of planning as a public service system before examining three more complex questions, and then the future.

Four objectives

To resolve conflict. Britain, especially the southern half, is a small and very crowded island with increasing population numbers. The pressure on land use is thus intense and a system is needed to resolve conflicts over land use. The private-sector land market provides one mechanism by pricing people out of certain areas, but this is inequitable and inefficient. Land use planning has provided an excellent arbitration service and most land use conflicts are resolved satisfactorily. There are vigorous debates and protest movements but they are peaceful and there have never been planning riots. Technically, planning has done the job well.

To be democratic and accountable. Planning has done a good technical job but only because the system is not very democratic or accountable. The two are linked. Technical jobs are best done by a few committed experts following strategic guidelines set down by politicians. Planning has not yet designed a camel, but some of planning's biggest errors have been where politicians have intervened. For example, central government allowing development on flood plains after local planners had refused permission. Nonetheless, planning must be accountable because the potential for corruption is very high. The balance between professional discretion and democratic accountability has by and large been a good one.

To reconcile private and public interest. Britain has once again become a free-market economy after a brief flirtation with socialism after the war but the public interest is still a key concept. In essence, any private development that interferes with the public interest should be refused. However, the public interest is a shadowy, elusive concept or a smokescreen which planners often conveniently use to mask the real reasons behind a decision. The public interest is thus a useful concept because it is indefinable and flexible. It is also politically popular because people can see that private wealth cannot buy everything, even a new house in the Green Belt.

To provide equity and fairness. Although planning does not discriminate overtly between rich and poor its decisions have had far-reaching implications for class divisions. The key case is over housing, where restrictive policies have forced up house prices in attractive areas way beyond the reach of locals, creating large upper middle-class areas in the countryside and in areas where little growth is allowed. In contrast, cheaper housing has been allowed in large 'estates' of incredibly dull design and remote from jobs and services. Planning has not created the sort of suburban environment envisaged by the new town seers but a cruel parody of it with cramped rooms, yards instead of gardens and cars overflowing on to roads. Planning has preserved the environment for a select few but created a shoddy second-rate environment for far too many.

Three big questions

Are the goals of planning fundamentally different from society and thus can planning only guide and shape change or can it make a real difference, for example by imposing greater sustainability? Planning exists because society perceives a need for it to provide the sort of services outlined above. If planning were to step outside these parameters it would find its existence to be endangered. Thus planning can only guide and shape change with the materials to hand. Contemporary society is based on a just-in-time service and manufacturing economy. This demands an environment based on long-distance road transport, edge-of-town employment areas at transport interchanges and residential areas slotted in where environmental constraints like good land, flooding or good landscape are not too high. This is not very sustainable but it is how the modern world works, and planning's job is to manage change to make it work as best as possible. So we are stuck in a vicious circle. Traffic grows because the economy grows, new roads like the M25 are built to reduce congestion. But businesses and people relocate to use the M25, the M25 becomes clogged, so another lane is added. More activities relocate and another lane is added and so on. This is the worst nightmare of 'predict and provide' planning, the snake swallowing its tail. But can we and do we want to get out of this circle? Day-to-day planning can deal only with reality, but strategic planners should offer us some alternatives.

More prosaically, do the delivery systems allow planning to work effectively and efficiently? In particular is the relationship between plan making and development control an effective one? Planning has helped to manage change although it tends to be one step behind because of the slowness inherent in preparing and approving plans. Another factor is that public investment does not anticipate demand but follows them. So roads are built only when congestion has reached a certain level. Development control is a remarkably efficient and flexible system, which has stood the test of time with only minor changes. The fact that plans are only one factor in development control is a major ingredient in this success story. Nonetheless, plans provide legitimacy and a framework within which to base consistent decisions. The relationship is structurally a very good one, akin to a well built car. If there are any problems with such a car they are due to the drivers and the environment in which they must drive, namely a society that is not yet ready to abandon short-term greed for long-term quality of life.

Do we have good enough evaluation systems and information to make a quantitative evaluation of planning or must we rely on inferential and qualitative judgements because at the end of the day planning is not an exact science? Most of the built environment has now been constructed under the planning system and it is an amazing point that no official evaluation of the system has been made. This is probably because it is an impossible task, largely because of the 'birth control' effect or what might have happened without planning. Furthermore, the complexity and tangled interrelationships in modern life make a quantitative evaluation virtually untenable. However, the modern world is awash with data and is over-dependent on electronic media and so potentially enough information is available. But the *Hitchhiker's Guide to the Galaxy*, which postulated that the universe was a huge computer to calculate the meaning of life, came up with the famous answer of 42. So no quantitative method can come up with answers to deep questions. Planning is ultimately about making things better and about value judgements, so it is appropriate that the ultimate evaluation should be a qualitative and inferential judgement, albeit one based on the widest possible evidence. There cannot, however, be *one* evaluation because of the diverse and conflicting roles that planning must play, and so there will always be qualifications.

The future

Should the system be modified and if so what are the alternatives? The system is essentially sound, with its twin pillars of plan making and development control. The main alternatives seem to be focusing on technical issues or to offer a vision of a more adventurous, proactive type of planning. These are discussed below before the book concludes with the government's latest reforms, which axiomatically provides its evaluation of the system.

Should planning focus even more on being a technical exercise based on key policy criteria or should planning break out from regulation and its narrow focus on balancing economic growth and conservation and use more of the Gilg–Selman spectrum? And if so should the environment be accorded greater importance in planning? Planning is most easily defensible as a technical reconciliation service managing change. It is thus tempting to recommend this route. However, to do so would be to accept the vicious circle of 'predict and provide' outlined above which might ultimately be unsustainable. For example, at what stage does south-east England become 'full up'? If planning is to change it must also change society. To do so it must offer an alternative vision. The new towns movement 100 years ago provided such a vision and an alternative to the high-density city. That vision was relatively easy, based as it was on new forms of road transport which enable suburbia and a leafy green environment based on clean new industries.

Today we have become locked into mobility partly because our suburban environments do not offer enough attractive space or enough social activity to satisfy us. Thus we fly to Barcelona for café society weekends or drive to National Parks to read our Sunday papers. In the working week we commute hundreds of miles and waste hours in traffic jams or on crowded trains. This is not life but mere existence. There must be a better way. But what is it and do we want to escape from our workcentric lives? I am afraid I do not have a vision as I am merely an academic writer skilled at summarising and commenting on research. What I do know and can suggest, though, is that the purposes of the system need to be redirected towards sustainability and mixed land uses, with a major reduction in car use. Ironically, this could be achieved by reversing the high-density housing mantra. Bigger houses and gardens would allow greater leisure use and food production. However, this would require major changes in attitudes to countryside preservation, the loss of much of lowland England and major changes in attitudes to food, away from ready-made meals and back towards home cooking. I do not see these changes as realistic, although I try to practise self-sufficiency in fresh vegetables and via log fires from sustainable felling in my modest garden.

The oft touted alternative is a move towards even higher densities and a society based on the celebrated European café culture. This will prove difficult, since surveys demonstrate that people want a free-standing house, with a garden, however small, and with space for probably two cars. Even though the result is rabbit hutches on postage stamps often 20 miles from work it is what people seem to want or need, with partners working in different places. The home is a base, the car is a mobile living room and the world is our oyster. Yet when more and more of us live like this we resemble lemmings rushing to the precipice albeit via traffic jams.

If we return to the vision of the new towns two principles stand out. First, land assembly by one authority. Second, self-contained communities with jobs, home and leisure in close proximity. It may not be possible to return to this ideal vision and indeed the new towns did not achieve it, but we ought to try again. Each region could build new towns to take its population forecast and the land could be bought by the public purse as in the 1950s. The idea is politically unrealistic in today's Public Finance Initiative climate when hospitals are rented from the private sector but nonetheless planning should offer alternatives to existing mantras. Publicly financed new towns built on land bought at existing use value could provide houses which reflected their building costs without the 40% land element. So my alternative is a bold one, based on big developments. We now turn to the government's proposals, which in contrast rely on incrementalism, albeit fairly radical incrementalism in terms of plan-making structures.

Summary

This section has used a hybrid approach to produce a set of related evaluations. The conclusion is that there cannot be one evaluation, given the complexity of the issues. It is concluded, though, that planning has been a *good thing* and has helped to maintain a fairly high-quality environment, notably in rural areas. The hybrid approach has been based on a checklist and the government has produced its own indicator checklist, as shown in Box 5.3. This is a more prosaic list than the list used above but it does return us to the gritty reality that planning is a technically dominated exercise.

Box 5.3   Best Value performance indicators for planning

1   Percentage of new homes built on previously developed land.
2   Planning cost per head of population.
3   The number of advertised departures from the plan as a percentage of permissions.
4   Percentage of applications determined in eight weeks.
5   Average time to determine applications.
6   Do you a have a development plan which was adopted in the last five years? If 'no':
   (a) Are there proposals on deposit for alteration or replacement and have you publicised a timetable for adopting these alterations or the replacement plan?
   (b) Does your plan contain a set of indicators and do you monitor your performance against these?
   (c) Has all supplementary planning policy followed the guidance in Planning Policy?
   (d) Do you provide for pre-application discussions with potential applicants?
   (e) Do you a have a publicised charter which sets targets for development control?
   (f) Is the percentage of appeals where the council's decision has been overturned lower than 40 per cent? Does your authority delegate more than 70 per cent of its applications to officers?
   (g) In the last year have you:
       (i)   Not had any planning costs awarded against you?
       (ii) Not had any adverse ombudsman's reports?
       (iii) Not had any court findings against you?
   (h) Does your council operate a one-stop shop service?
   (i) Have you implemented a policy for ensuring that different groups have equal access to the planning process, including the provision of advice in ethnic minority languages?

Source: Cullingworth and Nadin (2002)

THE FUTURE AS A MIRROR OF THE PAST

When we look to the future we base our vision on an evaluation of where we have come from. Looking to the future thus provides us with one final evaluation method. This final section is divided into two main sections. First, background trends. Second, proposals for reform from various organisations, government proposals for reform made in 2001 and 2002 which led to a major new Planning Act in 2004, and finally my evaluation and manifesto for the future. Before this analysis it is essential to remind ourselves that forecasting the future is fraught with difficulties and planning will always have to accept that its forecasts can be only guides, not predictions.

Background trends

This section begins with some key points about what the future may hold, namely projected changes in population and projected changes in lifestyles. Forecasts in population change have projected a static native population, but that does not mean a static demand for development. First, household sizes have been declining as young people leave home earlier, divorce has led to more single-person households in middle age, and old people are living longer but often alone. However, these trends may not continue as expected. First, student tuition fees and debts may encourage more young people to live at home, a trend that has already begun, according to new data on people living with parents into their late twenties. Second, divorce rates have levelled out and even fallen. Third, rising house prices have meant that it is more difficult for single people to buy a house. However, old people are living longer and an ageing population will cause social and economic problems in the future as they become infirm and need care. A new feature is rapidly rising immigration, both legal and illegal, and new forecasts are showing major increases in Britain's population for the next 20 years as young immigrants are brought in to keep the economy going as the native UK population ages. However, as Chapter 1 has shown, population forecasts have been wildly inaccurate in the past and forecasts only five years apart can show variations of over a million households, so caution needs to be exercised.

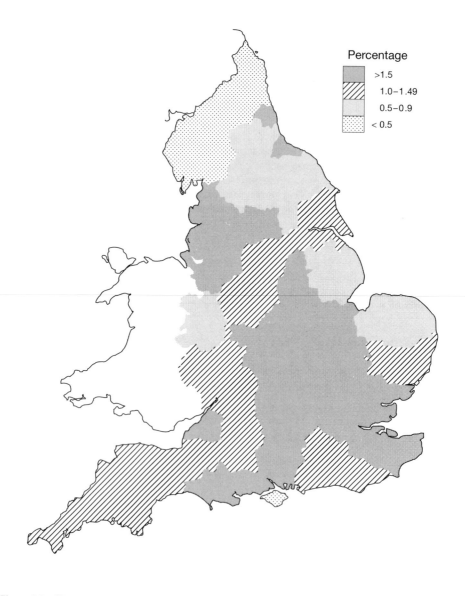

Figure 5.7   The projected rate of change to urban uses, 1991–2016. *Source*: adapted from Bibby and Shepherd (1997)

Most recent plans have focused on household projections in 1996, which forecast an extra 4.4 million households by 2016. Figure 5.7 shows where most of the 4.4 million extra households are expected to be accommodated in terms of the percentage of the county area projected to change from rural to urban use. It shows that once again it is central southern England that will continue to bear the brunt of change but with significant growth in the North West. However, when the projections are recast as a percentage of the 1991 area under urban use the picture changes radically. Indeed, Figure 5.8 shows that an arc from Cornwall to Suffolk will experience the most rapid rate of urbanisation, with outliers in the West Midlands and the North East.

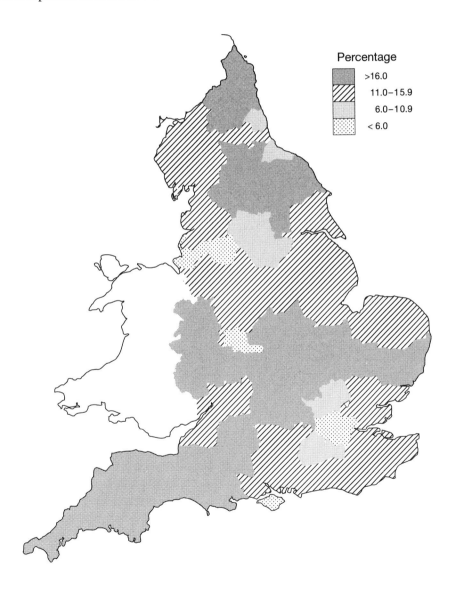

Figure 5.8 The projected rate of urban growth, 1991–2016. *Source*: adapted from Bibby and Shepherd (1997)

The most recent projections have reflected the increase in migration, and the population of the UK is now expected to rise from 59 million in 2001 to 64 million by 2021 (Office of National Statistics, 2002). This number will depend on fluctuating attitudes to immigration both in the United Kingdom and elsewhere as international events (reduction of tyrannical governments) and the impact of the enlargement of the European Union change the demand for entry to the United Kingdom. Within the United Kingdom, immigration is now regarded as a good thing officially in order to fill labour shortages in lower-paid sectors of the economy and public services like the NHS. Meanwhile the long-awaited results of the 2001 census have yet to find their way into detailed regional and local forecasts.

The second change, changing lifestyles, is even harder to project. First, new technologies are reducing even further the need to work in a static central workplace or to shop in fixed shops. So-called e-working and e-retailing has not yet had the impact it was expected to but it may yet allow us to return to a more home-based lifestyle in which we spend more time at home and in local surroundings. Indeed, the first draft of this book was written on a laptop 135 miles from home, during a family illness, and the second and third drafts in my study at home. If home working expands significantly we will want bigger houses and gardens, and current trends to infill the suburbs may not be acceptable. The main debate is between those who espouse the suburban 'idyll', notably the Town and Country Planning Association (1999), and those who extol the virtues of Continental café societies like Barcelona (Urban Task Force, 1999). Much will depend on how we want to share our lives with each other and the degree to which traditional nuclear families will be formed.

Our changing attitudes are plotted regularly by the British Social Attitudes Survey and it has consistently shown a significant mismatch between growing environmental concern and our actual behaviour. For example, nearly 80% of respondents want the countryside and Green Belts to be protected, and traffic to be cut even at the expense of jobs (Christie and Jarvis, 1999). In contrast, surveys for the Countryside Commission show that most people want a house in the countryside, which would lead to unsustainable traffic increases. We want the ends but not the means and few politicians have the courage to interfere with our freedom to drive. Indeed, private mobility is seen as a right rather than a privilege. The contradiction between one individual's desire for a pleasant free-standing house and for a traffic-free road and the reality of a small crowded island lies at the heart of planning's predicament.

Proposals for reform

Turning to proposals for the future, these can be divided up into those by academics and lobby groups and those by official organisations and the government.

Non-governmental proposals

Starting with academic proposals, Rydin (1998) has argued that planning should set out specific goals that are based on a consensus among key actors, following the model set out by the Biodiversity Convention at the 1992 Rio Summit, which suggests that planning strategies should be limited to:

1   Clear goals and specific targets.
2   Potential mechanisms for achieving these targets.

3   The lead agencies involved.
4   The resource implications of the strategy.

Carmona *et al.* (2003) have called for a new culture in planning, which should have eight essential features. First, efficient in decision making. Second, equitable in processes and outcomes. Third, providing co-ordinated policy responses to complex problems. Fourth, sensitive to change. Fifth, capable of delivering predictable high-quality outcomes. Sixth, ethical and accountable. Seventh, visionary. Eighth, effective at delivering change.

An example of reforms by a lobby group has been provided by the Town and Country Planning Association (1999), which set out seven principles for guiding reform. Among them was the principle that the overall purpose of planning is to manage change, a view endorsed by Gilg and Kelly (2000) in their paper on planning for the new millennium. But because this process is not value-free planning must be for people and must be more inclusive. The Association argued that a complete change in culture was needed. One way in which it might be achieved is through the way in which planners are educated at university. However, Poxon (2001) from a series of focus groups found that there were far more questions than answers about how training should be modified, except that it should be directed to higher-level skills like evaluation and analysis.

An example of proposals for change came from the Royal Commission on Environmental Pollution (2002). The proposals were based on a central premise of this book, that economic and social objectives have to be achieved in ways that safeguard and enhance the environment. It then argued that town and country planning would continue to be of central importance in ensuring that balance, but that there were five key problems with the present arrangements:

1   Planning does not provide an integrated, accountable and transparent way of setting and achieving environmental goals at different levels.
2   There is a complex variety of legislation and a multiplicity of often overlapping and sometimes conflicting plans and strategies across the environmental field (see Table 5.7).
3   Responsibilities are divided between the UK government, the devolved administrations, local authorities and specialist agencies.
4   There is no succinct statement of priority objectives for the environment.
5   Rather than being seen as an essential instrument of environmental planning, town and country planning has come to be seen as a bureaucratic structure of doubtful relevance.

The commission concurred with the widespread view that the system was in need of reform, in spite of its acceptance that without the system widespread damage to the environment would have occurred. Thus the system had achieved a lot, albeit negatively, but could do so much better.

In order to do better, the commission argued, planning needed a clear statutory purpose. First, the commission proposed the following statutory purpose for planning: 'To facilitate the achievement of legitimate economic and social goals whilst ensuring that the quality of the environment is safeguarded and, wherever appropriate, enhanced' (p. 4). In addition, the commission proposed six areas for increasing efficiency and effectiveness in planning:

1   Clearer policies and objectives for the environment.

Table 5.7 Environment or environment-related plans in England

	Government office	Planning authorities/ regional planning body	Other local authority plans	Specialist agencies
Regional	1 Renewable Energy 2 Rural Development Programme	1 Regional Planning Guidance 2 Regional Transport Strategy 3 Regional Sustainability Framework		1 Economic Strategy (RDA) 2 Regional Look Forward (EA) 3 Water Resources Strategy (EA) 4 Biodiversity Audit (EN)
Sub-regional		1 Structure Plan 2 Waste Plan 3 Minerals Plan 4 Supplementary guidance on specific topics	1 Community Strategy 2 Local Transport Plan 3 Local Agenda 21 strategy 4 Municipal Waste	1 Biodiversty Action Plan (BP) 2 Shoreline Management Plan (EA/LA)
Local		1 District-wide Local Plan 2 Supplementary planning guidance on specific topics	1 Community Strategy 2 Air Quality Management Plan 3 Local Agenda 21 strategy	1 Local Environment Agency Plan (EA) 2 Coastal Habitat Management Plan (EN/EA/LA) 3 Nitrate Vulnerable Zones (DEFRA)

DEFRA Department for Food and Rural Affairs, *EA* Environment Agency, *EN* English Nature, *LA* Local authority, *RDA* Rural Development Agency.

Source: Royal Commission (2002)

2   Statutory recognition of the central role of town and country planning in protecting and enhancing the environment.
3   Rationalising the overall system of environmental planning by introducing integrated spatial strategies covering all aspects of sustainable development.
4   Ensuring that strategies cover all forms of land use.
5   Much improved availability of information about the environment.
6   Further steps to involve a wider range of people in decisions about setting and achieving environmental goals.

In particular, the commission argued that the long-standing presumption in favour of development ought no longer to apply and that legislation should stipulate key aspects of the environment and natural resources as material considerations which should be taken into account in considering all planning applications. In terms of plan making, the commission recommended the introduction of integrated spatial strategies covering the atmosphere, water and land use for at least 25 years ahead

This goes further than the traditional role of balancing conflicting aims, which the commission believed to be too facile. Instead the commission proposed that the goal of protecting and enhancing the environment must be 'fundamental' (p. 38). In order to achieve this goal the commission saw a considerable role for economic and financial incentives but as complementary to regulation, not as a replacement. Indeed, they argue that direct regulation of land use remains a vital instrument for promoting sustainable development and safeguarding environmental sustainability.

These examples of academic and lobby group proposals are, however, idealistic and in the real world of power politics the only proposals that really matter are those that come from the government, although these may of course have been influenced during their gestation by lobby groups and academics. The most notable proposal to change the planning system in recent years, and indeed a very radical document in terms of plan making, was the Planning Green Paper for England of December 2001 (DTLR, 2001) and the subsequent new Act of 2004.

Government proposals: the UK Sustainable Development Strategy, the 2001 Planning Green Paper and the 2004 Planning Act

The UK Sustainable Development Strategy (Royal Commission, 2002) published in 1999 identified the following *future priorities* for the United Kingdom:

1   More investment in people and equipment for a competitive economy.
2   Reducing the level of social exclusion.
3   Promoting a transport system which provides choice and also minimises environmental harm and reduces congestion.
4   Improving the larger towns and cities to make them better places to live and work.
5   Directing development and promoting agricultural practices to protect and enhance the countryside and wildlife.
6   Improving energy efficiency and tackling waste.
7   Working with others to achieve sustainable development internationally.

It also identifies ten *guiding principles* for government policies:

1   Putting people at the centre.
2   Taking a long-term perspective.
3   Taking account of costs and benefits.
4   Creating an open and supportive economic system.
5   Combating poverty and social exclusion.
6   Respecting environmental limits.
7   The precautionary principle.
8   Using scientific knowledge.
9   Transparency, information, participation and access to justice.
10   Making the polluter pay.

The 2001 Green Paper (available on government Web sites) began with an analysis of why changes were needed based on the weaknesses of the current system, which were argued to be:

1   Plans have been prepared too slowly and are too detailed.

2   Plans contain contradictory policies, which are not prioritised.
3   Conflict between plans in the hierarchy, notably between RPG and county councils.
4   Also tensions between RPG and RDA plans.
5   Plans have not engaged public interest.
6   Plan-making areas are not relevant to everyday life.
7   Plans do not provide good guides for development control.
8   Not enough commitment to sustainability.
9   Forecasting based on trends rather than attempts to alter the future.

The objectives of the Green Paper were to ensure that planning has a vital role to play in the quality of life, notably in encouraging urban regeneration, conserving greenfield land and delivering sustainable development. But at the same time planning must deliver speedy decisions that do not frustrate business and the planning system must accommodate change, not just resist and stifle it. Thus the rhetoric was dominated by the need for speed, efficiency and certainty, so: 'There will be a fundamental change in planning so that it works much better for business' (para. 2.10).

The main proposals in the Green Paper were directed at plan making but within four main overall aims:

1   Simplify the hierarchy and clarify the relationships within it.
2   Produce shorter, better focused plans at the local level which can be adapted and revised more quickly.
3   Engage the community more closely.
4   Improve integration with other local strategies and plans.

For plan making the framework shown in Figure 5.9 was proposed. In this framework, Structure, Local and Unitary Development Plans are to be abolished. Also Regional Planning Guidance would be replaced by Regional Spatial Strategies (RSSs). The RSS would be a statutory plan to be followed by all plans down the hierarchy. The RSS should outline specific regional or sub-regional policies; provide the broad outlines of major development proposals; set targets and indicators; and cross-refer rather than repeat national policy. The RSS should also provide the longer-term planning framework for the RDAs. The RSS would be prepared by Regional Planning Bodies which should be representative of key regional interests. Sub-regional strategies may be prepared for: major conurbations; where the planning of major towns and cities may raise strategic issues which can be resolved only on a joint basis by neighbouring local authorities; and to develop strategies which straddle regional or county boundaries.

A subsequent White Paper in May 2002 set out proposals for directly elected Regional Assemblies which would take over the regional planning role if regional referendums were in favour of such assemblies. If such assemblies were set up existing local authorities in the region would almost certainly be abolished to be replaced by unitary districts.

At the local level Local Development Frameworks (LDFs) would be prepared in tandem with the recently introduced Community Strategies. Community Strategies have four key components, most notably a long-term vision and an action plan for shorter-term priorities. Working with the Community Strategy, the LDF should consist of:

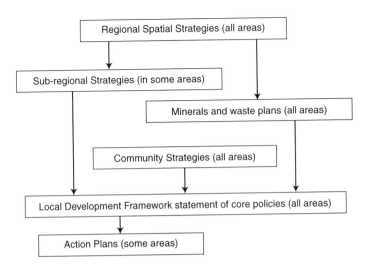

Figure 5.9 Revised plan-making system envisaged by the 2001 Green Paper. *Source:* author

1   A statement of core policies setting out the authority's vision and strategy to be applied in controlling development.
2   More detailed action plans for smaller local areas of change such as urban extensions, town centres and neighbourhoods undergoing renewal.
3   A map showing the areas of change for which action plans are to be prepared.
4   Existing designations such as Conservation Areas, National Park/AONB management plans to remain.

LDFs would be prepared by the relevant district, unitary or National Park Authority and should be prepared in *months* rather than years. The statement of core policies should be continuously updated and should be consistent with national and regional policies. The LDF should contain a Statement of Community Involvement, setting out how the community will be involved with the continuing review of the LDF, and commenting on significant planning applications. The LDF must contain a Sustainability Appraisal following the EU Directive on Strategic Environmental Assessment.

The LDF may also identify areas for Action Plans. Possible Action Plans could include:

1   Area Master Plans for renewal or new build.
2   Neighbourhood and village plans setting out how the distinctive character of an area is to be preserved, the location of any new development, design standards, and how to identify key services and facilities.
3   Design Statements setting out standards for an area or type of development.
4   Site development briefs setting out detailed guidance on how a particular site should be developed.
5) Action Plans may also be prepared for Green Belts, housing allocations, specific proposals for major developments and safeguarding land for transport or other purposes.

In contrast to these radical proposals for plan making the proposals for development control were concentrated around detailed changes to make the system quicker, more predictable and more transparent, mainly through the application of checklists for both applicants and planners. Checklists have been endorsed by Kelly *et al.* (2004) but only if they force planners and councillors to be more aware of the environmental consequences.

Early comments on the proposals were centred on the bipolar nature of the plan–making arrangements, focusing at the regional and the local level, thus cutting out the structure plan or county level. In addition two hidden agendas were discerned. First, an EU agenda based on developing a Europe of the Regions, hence the emphasis on regional plans. Second, a Tony Blair agenda for modernising Britain based on local communities, hence the emphasis on Community Strategies and Local Development Frameworks at the expense of traditional counties.

The Green Paper was met with nearly universal disapproval from the planning community both because of its proposal to cut structure plans and because of its emphasis on helping business. In spite of these criticisms the government published its response in July 2002 (Lock, 2002), which despite the criticism contained most of the proposals in the Green Paper. One significant omission was a proposal to give Parliament the power to decide on major applications like airports rather than the public inquiry system. One major new proposal was to include a statutory purpose for planning, remedying a major problem for evaluation since 1947. In December 2002 the Planning and Compulsory Purchase Bill was introduced to Parliament and after much delay was given the royal assent in May 2004. The main changes introduced by the Planning and Compulsory Purchase Act 2004, which made sustainable development the main purpose of planning, are:

1   Structure Plans to be abolished.
2   Regional Spatial Strategies (RSSs) to be drawn up by Regional Chambers (Assemblies).
3   Local Development Schemes to set out procedures for preparing Local Development Frameworks.
4   Local Plans (collectively a Local Development Framework) to be compatible with RSS and national planning policies and to demonstrate community involvement.
5   Local Development Documents will have core policies, development proposals and a clear map and Local Development Orders can extend permitted development rights under the GPDO in order to speed up the system and provide more certainty for developers and communities.
6   Introduction of Business Planning Zones, whose location has been determined by RSSs, within which planning permission should be a formality as long as the proposed development is in accordance with tightly defined parameters.
7   Planning permission reduced from five to three years.

At the same time the government published a draft Planning Policy Statement 1 (PPS1) to replace and update PPG1 in the light of the new Act. It began by setting out a threefold vision for planning with three key themes:

Threefold vision
1   Planning has a critical role in delivering the government's wider macroeconomic, social and environmental objectives.
2   Planning should aspire to make places better for people and deliver development where communities need it and where it is sustainable.
3   Good planning should seek to manage development, not simply control it.

Three key themes
1   Sustainable development.
2   Spatial planning approach.
3   Community involvement.

PPS1 also stresses a transition to spatial planning from land use planning in order to integrate policies for the development of land with other policies and programmes that influence the nature of places and how they function. Spatial planning should also help to deliver policy in four key areas. First, the need for planning authorities to take an approach based on the four aims of economic development, social inclusion, environmental protection and the prudent use of resources. Second, the need for positive planning to achieve sustainable development objectives and proactive management of development rather than simply regulation and control. Third, the need for plans to set clear visions for communities and help to integrate the wide range of activities relating to development and regeneration. Fourth, the need for the planning system to be transparent, accessible and accountable, and to actively promote participation and involvement.

Early criticisms by Rydin and Deegan (2004) of PPS1 welcomed the wider remit and vision for planning but feared that the new Act bestow the powers to actually provide holistic and sustainability-driven spatial planning. More fundamentally, Kelly *et al.* (2004) have criticised the definition of sustainability which mentions high and sustained rates of economic growth in the context of increasing pressures on basic resources of water and fossil fuels.

Reform proposals for Scotland and Wales were also made in 2001 and 2002.

PERSONAL POSTSCRIPT

My evaluation

My evaluation, based on 35 years of professional experience, 57 years of life and several months spread over 3 years writing this book, is that planning has been a *good thing* and that we desperately need it, as is shown any time one travels around Britain, as shown in Box 5.4. But there are six main qualifications:

1   Planning has prevented many bad things but it has not always managed to stop damage, notably with regard to public developments in protected areas.
2   It has had a number of side effects, the most serious of which has been inflationary land values and related house price inflation and shortages, notably in the South East.
3   It has worked by zoning land but this has led to the other serious side effect, the separation of land use activities, leading to over-dependence on car use, long travel distances and severe problems of congestion.
4   It is most effective (a) at the local level; (b) in partnership; (c) over short time horizons.
5   It has had little effect on agriculture and forestry even though these land uses have undergone radical change, usually for the worse, with dull monocultural landscapes largely devoid of wildlife taking over from a rich and diverse heritage. Villages have suffered from the addition of alien housing estates.
6   It has led to committee-style unadventurous landscapes, notably Britain's appalling suburbs with their banal and derivative architecture from which people have to travel everywhere because of the lack of work, shopping and leisure facilities.

Box 5.4 Reflections on planning on a weekend away in July 2004

While grappling with the final draft of this text I had a long weekend away from Exeter. First, on Friday morning my wife and I drove up the M5 motorway to visit our son in Bristol. The motorway was busy with early holiday traffic but free-flowing, though in a stop–start way, as slow-moving vehicles temporarily slowed the middle lane and thus the outer lane. This behaviour exemplifies the planning dilemma of liberty or control. On a controlled motorway, vehicles that take several hundred metres or longer to overtake would be fined or banned and the motorway would flow much more freely, but would this be too draconian? But where is everybody going?

Our son has just moved house in Bristol from a Victorian brownstone terrace to a converted hotel development in a terrace of elegant Georgian town houses built above the Avon gorge and with a stunning view of the Clifton bridge and green fields beyond. We approached his penthouse via a jumble of lanes enclosed by mature stone walls, overhanging tress and homely pubs. Cows to the west, cafés to the east, the planning idyll? A wonderful urban village, enhanced by the mix of chic restaurants, boutiques and everyday shops that provide a model for the so-called café society planners to eulogise about. But two thoughts: would the Clifton bridge be built today and why is there little development to the west of it? The Green Belt, of course.

Instead Bristol has sprawled to the north-east and as we left Bristol we could see the Americana horror of Cribb's Causeway, an ants' heap of hideous grey boxes betopped by garish signs proclaiming the chain stores or alleged restaurants to be found there. And yet people drive from Exeter to shop and eat there, and Exeter's 1950s' town centre is to be demolished to make way for an inner-city complex of similar businesses in order to compete. To middle-class Clifton eyes this is strange behaviour, but out-of-town shopping is favoured by large numbers and by global mega-corporations. Planning here has given in to economic forces and given people what they vote for with their feet. But is this a lowest common denominator?

At least Cribb's Causeway is efficient, but our drive through Gloucester is tediously slow as we negotiate the continual repair of our decaying Victorian infrastructure, via a two-mile strip of assorted small businesses in cheap and nasty buildings of extremely varied styles. The environment is tawdry and frankly depressing but there is evidence of a vibrant local community and a thriving local trade economy. Gloucester works but in a totally different way from Cribb's Causeway.

Eventually we arrive at my mother's house in a village four miles to the north of Hereford. Once a small village astride a minor road, it grew enormously in the 1970s and 1980s with hundreds of small houses. It is entirely a dormitory estate, and as I take an evening stroll a constant stream of cars with just one occupant create a jumble of two, sometimes three cars outside each house. Indeed, cars take up as much space as the inadequate house and front garden space. Why did planners allow these dormitory estates, based on energy-costly commuting, and why do I find it depressing when the inhabitants believe themselves to be living happily in the country? Is it my hamlet upbringing, my middle-class spectacles or ecological concern?

On Saturday morning we visit Hereford city centre, once an earthy market town but now dominated by light industry. Even in the pedestrianised market square, charity shops prevail and the spectre of the global out-of-town attractions hangs heavy even in a city thirty miles from a major motorway and fifty miles from a big city. Finally we play golf on two commercial courses. One surrounded by horticulture with the plastic sheet menace, largely outside planning control, and one a hotel course with a Travelodge-style hotel sitting uneasily beside an old mansion.

So what do we conclude from this odyssey? Overwhelmingly that our society and economy are dominated by movement. Indeed, we have become obsessed with travel. The northbound M5 on Sunday evening is, as ever, clogged by returning second-home owners. What is so wrong with our home environment that we feel this compulsive need to get away all the time? Our television channels are saturated with house hunting and travel programmes. We fantasise about other places, because planning has created too many banal spaces. We need to re-create Clifton.

So let's cut the need to travel by creating vibrant local communities.

My manifesto

Finally, my manifesto for the future would involve the following:

1   Planning to be led by sustainable development principles in which development, except for necessary organic growth of settlements or businesses, would be prohibited in most areas. Development would be allowed only in areas where it could contribute to environmental improvement via planning gain and would be identified by:

2   Regional Planning Assemblies producing regional strategies to identify suitable areas for organic growth and one or two major new settlements (see 5e).

3   Unitary City Regions producing Area-wide Local Plans.

4   Development control to be plan-led.

5   In terms of the core land uses the following principles would apply:

   (a)   City centres should encourage repopulation by insisting that growth should be accommodated above commercial uses.

   (b)   Inner-city regeneration areas should insist on mixed uses.

   (c)   Suburbs should be encouraged to become multi-use, with urban village centres providing work, shops and leisure.

   (d)   No more housing should be allowed in villages except for proven local need to house workers employed in the village or near by.

   (e)   New self-contained settlements up to 100,000 people should be allowed to accommodate regional growth and change.

Further reading
in addition to references already highlighted in text and illustrative material

Now that you have grasped the key issues in the evaluation of planning you will be ready to enjoy Nigel Taylor's excellent, albeit mistitled, book *Urban Planning Theory since 1945*. The book is in fact an historical account of planning and is not about urban planning theory. His key argument is that planning is a form of social action directed at shaping the physical environment. Planning is thus a reflection of the society it finds itself in but if it is to have real credibility it has to provide a vision of how planning might improve society by providing better day-to-day places. Other useful texts to widen your appreciation are:

Bramley, G., Bartlett, W. and Lambert, C. (1995) *Planning, the Market and Private House Building: The Local Supply Response*, London, UCL Press.

Booth, P. (1996) *Controlling Development: Certainty and Discretion in Europe, the USA and Hong Kong*, London, UCL Press.

Selman, P. (1996) *Local Sustainability: Managing and Planning Ecologically Sound Places*, London, Chapman.

Selman, P. (2000) *Environmental Planning*, London, Sage.

Ward, S. (2004) *Planning and Urban Change*, London: Sage. This book is an ideal companion to this chapter because it provides many detailed and illustrated examples of the physical impact of planning.

Most organisations with a stake in planning will have part of their Web site devoted to their analysis of planning and their ideas for the future. Divide these into official, respected and fringe groups before evaluating what they have to say.

Bibliography

A note on parliamentary publications. White and some Green Papers are called Command Papers (by Command of Her Majesty) and are numbered 1–1000 with a prefix of Cm, Cmd or Cmnd, depending on the period. House of Commons and House of Lords papers are classified by the parliamentary year, e.g. 97–98, and in ascending number through the year. So HC 54 (97–98) is the fifty-fourth paper of the session in 1997–98. Bills are given similar numbers. Acts of Parliament are given a chapter number by ascending number for the parliamentary year.

Abram, S., Murdoch, J. and Marsden, T. (1998) Planning by numbers: migration and statistical governance, in *Migration into Rural Areas*, edited by P. Boyle and K. Halfacree, Chichester, Wiley, pp. 236–51.

Adams, D. (2004) The changing regulatory environment for speculative housebuilding and the construction of core competencies for brownfield development, *Environment and Planning A*, 36, pp. 601–24.

Adams, D. and May, H. (1990) Land ownership and land use planning, *The Planner*, 76, 38, pp. 11–14.

Adams, D. and May, H. (1992) The role of landowners in the preparation of statutory local plans, *Town Planning Review*, 63, 297–323.

Adams, D., Disbery, A., Hutchinson, N. and Munjoma, T. (2001) Ownership constraints to brownfield development, *Environment and Planning A*, 33, pp. 453–77.

Alder, J. (1989) *Development Control*, London, Sweet and Maxwell.

Alexander, E. (1998) Doing the impossible: notes for a general theory of planning, *Environment and Planning B: Planning and Design*, 25, pp. 667–80.

Allmendinger, P. and Tewdwr-Jones, M. (2000a) Spatial dimensions and institutional uncertainties of planning and the new 'regionalism', *Environment and Planning C*, 18, pp. 711–26.

Allmendinger, P. and Tewdwr-Jones, M. (2000b) New Labour, new planning? *Urban Studies*, 8, pp. 1397–402.

Allmendinger, P., Tewdwr-Jones, M. and Morphet, J. (2004) New-order planning and local government reforms, *Town and Country Planning*, 72, pp. 274–7.

Anderson, M. (1981) Planning policies and development control in the Sussex Downs AONB, *Town Planning Review*, 52, pp. 5–52.

Audit Commission (1992) *Building in Quality: a Study of Development Control*, London, HMSO.

Baker, M. (1995) Return to the regions, *Town and Country Planning*, 64, pp. 280–2.

Baker, M. (1999) Intervention or interference? Central government involvement in the plan-making process, *Town Planning Review*, 70, pp. 1–4.

Barker, K. (2004) *Review of Housing Supply: Delivering Stability: Serving our Future Housing Needs: Final Report,* London, HM Treasury.

Barlow report (1940) *Report of the Royal Commission on the Distribution of the Industrial Population*, Cmd 6153, London, HMSO.

Barrass, R. (1979) The first ten years of English structure planning, *Planning Outlook*, 22, pp. 19–23.

Bedford, T., Clark, J. and Harrison, C. (2002) Limits to new public participation practices in local land use planning, *Town Planning Review*, 73, pp. 311–31.

Bennett, G. (1981) Cartoon depicting evolving design styles in post-war Britain, *Planning*, 13 March, p. 2.

Best, R. (1977) Agricultural land loss: myth or reality? *The Planner*, January, pp. 15–16.

Best, R. and Champion, A. (1970) Regional conversion of agricultural land to urban use 1945–67, *Transactions of the Institute of British Geographers*, 49, pp. 15–31.

Bibby, P. and Shepherd, J. (1997) Projecting rates of urbanisation in England 1991–2016, *Town Planning Review*, 68, pp. 93–124.

Bibby, P. and Shepherd, J. (2001), Refocusing national targets for accommodating housing on brownfield sites, *Town and Country Planning*, 70, 12, pp. 327–31.

Bingham, M. (2001) Policy utilisation in planning control: planning appeals in England's 'plan-led' system, *Town Planning Review*, 72, pp. 321–40.

Bishop, K., Tewdwr-Jones, M. and Wilkinson, D. (2000) From spatial to local: the impact of the European Union on local authority planning in the UK, *Journal of Environmental Planning and Management*, 43, pp. 309–34.

Blowers, A. (2000) Ecological and political modernisation: the challenge for planning, *Town Planning Review*, 71, pp. 371–93.

Booth, P. (2003) *Planning by Consent: The Origins and Nature of British Development Control*, London, Routledge.

Bracken, I. and Hume, D. (1982) Forecasting techniques in Structure Plans, *Town Planning Review*, 52, pp. 375–89.

Bramley, G. (1993) Land use planning and the housing market in Britain, *Environment and Planning A*, 25, pp. 1021–51.

Breheny, M. (1995) The compact city and transport energy consumption, *New Transactions of the Institute of British Geographers*, 20, pp. 81–101.

Breheny, M. (1998) Success in reusing urban land, *Town and Country Planning*, 67, pp. 24–5.

Brindley, T., Rydin, Y. and Stoker, G. (1988) *Remaking Planning: The Politics of Urban Change in the Thatcher Years*, London, Routledge.

Brindley, T., Rydin, Y. and Stoker, G. (1996) *Remaking Planning: The Politics of Urban Change*, London, Routledge.

Brotherton, I. (1982) Development pressure and control in the National Parks 1966–81, *Town Planning Review*, 53, pp. 429–59.

Brotherton, I. (1984) A response to McNamara and Healey, *Town Planning Review*, 55, pp. 97–101.

Brotherton, I. (1992a) On the control of development by planning authorities, *Environment and Planning B*, 19, pp. 465–78.

Brotherton, I. (1992b) On the quantity and quality of planning applications, *Environment and Planning B*, 19, pp. 337–57.

Bruff, G. and Wood, A. (2000a) Local sustainable development: land-use planning's contribution to modern local government, *Journal of Environmental Planning and Management*, 43, pp. 519–39.

Bruff, G. and Wood, A. (2000b) Making sense of sustainable development: politicians, professionals, and policies in local planning, *Environment and Planning C*, 18, pp. 593–607.

Bruton, M. and Nicholson, D. (1985) Strategic land use planning and the British development plan system, *Town Planning Review*, 56, pp. 21–41.

Bruton, M. and Nicholson, D. (1987a) A future for development plans, *Journal of Planning and Environment Law*, October, pp. 687–703.

Bruton, M. and Nicholson, D. (1987b) *Local Planning in Practice*, London, Hutchinson.

Buller, H. and Hoggart, K. (1986) Non-decision making and community power: residential development control in rural areas, *Progress in Planning*, 25, pp. 135–203.

Bunnell, G. (1995) Planning gain in theory and practice: negotiation or agreements in Cambridgeshire, *Progress in Planning*, 44, pp. 1–113.

Burton, E. (2002) Measuring urban compactness in UK towns and cities, *Environment and Planning B*, 29, pp. 219–50.

Cambridge Architectural Research (2004) *Housing Futures: Informed Public Opinion*, York, Joseph Rowntree Foundation.

Campbell, H. and Marshall, R. (2000) Moral obligations, planning, and the public interest: a commentary on current British practice, *Environment and Planning B*, 27, pp. 297–312.

Campbell, H., Ellis, H. and Henneberry, J. (2000) Planning obligations, planning practice and land-use outcomes, *Environment and Planning B*, 27, pp. 759–75.

Campbell, S. (1996) Green cities, growing cities, just cities? Urban planning and the contradictions of sustainable development, *Journal of the American Planning Association*, 62, pp. 296–312.

Carmona, M. (2003) Planning indicators in England: a top-down evolutionary tale, *Built Environment*, 29, pp. 355–74.

Carmona, M. and Sieh, L. (2004) *Measuring Quality in Planning: Managing the Performance Process*, London, Spon.

Carmona, M., Carmona, S. and Gallent, N. (2003) *Delivering New Homes: Processes, Planners and Providers*, London, Routledge.

Carter, J., Wood, C. and Baker, M. (2003) Structure Plan appraisal and the SEA directive, *Town Planning Review*, 74, pp. 395–422.

Champion, T. (1989) Quality of life considerations and the prospects for growth, *The Planner*, 75, pp. 16–20.

Champion, T. (1993) A decade of regional and local population change, Census 91, *Town and Country Planning*, 62, pp. 42–50.

Champion, T., Coombes, M. and Shaw, T. (1983) A new definition of cities: population change 1971–81, *Town and Country Planning*, 52, pp. 305–7.

Cherry, G. (1974) *The Evolution of British Town Planning*, London, Intertext.

Cherry, G. (1979) The town planning movement and the late Victorian city, *New Transactions of the Institute of British Geographers*, 4, pp. 306–19.

Cherry, G. (1996) *Town Planning in Britain since 1900*, Oxford, Blackwell.

Christie, I. and Jarvis, L. (1999) Rural spaces and urban jams, in *British Social Attitudes: the 16th Report*, edited by R. Jowell, J. Curtice, A. Park and K. Thomson, Aldershot, Ashgate, pp. 113–34.

Claydon, J. (1996) Negotiations in planning, in *Implementing Town Planning*, edited by C. Greed, Harlow, Longman, pp. 110–20.

Cloke, P. (1987) Policy and implementation decisions, in *Rural Planning Policy into Action*, edited by P. Cloke, London, Harper and Row, pp. 19–34.

Cloke, P. (1996) Housing in the open countryside: irresponsible planning in rural Wales, *Town Planning Review*, 67, pp. 291–308.

Cloke, P. and Shaw, D. (1983) Rural settlement policies in Structure Plans, *Town Planning Review*, 54, pp. 338–54.

Cm 3016 (1995) *Rural England: A Nation committed to a Living Countryside*, London, HMSO.

Cm 3885 (1998) *Planning for the Communities of the Future*, London, HMSO.

Cmnd 4039 (1969) *Reorganisation of Local Government in England and Wales*, chairman Redcliffe Maud, London, HMSO.

Coates, D. (1992) Affordable housing: the private sector perspective, *The Planner, Proceedings of the Town and Country Planning Summer School*, 27 November, pp. 57–9.

Coleman, A. (1976) Is planning necessary? *Geographical Journal*, 142, pp. 411–37.

Corkindale, J. (1999) Land development in the United Kingdom: private property rights and public policy objectives, *Environment and Planning A*, 31, pp. 2053–70.

Counsell, D. (1999a) Making sustainable development operational, *Town and Country Planning*, 68, pp. 131–3.

Counsell, D. (1999b) Sustainable development and structure plans in England and Wales: operationalising the themes and principles, *Journal of Environmental Planning and Management*, 42, pp. 45–61.

Counsell, D. (2001) A regional approach to sustainable urban form? *Town and Country Planning*, 70, pp. 332–5.

Counsell, D. and Haughton, G. (2003) Regional planning tensions: planning for economic growth and sustainable development in two contrasting English regions, *Environment and Planning C*, 21, pp. 225–39.

CPRE (Campaign to Protect Rural England) (2004) *Prescott's Greenfield Hitlist*, London, The Campaign.

Crook, A. and Whitehead, C. (2002) Social housing and planning gain: is this an appropriate way of providing affordable housing? *Environment and Planning A*, 34, pp. 1259–79.

Cross, D. and Bristow, M. (1983) *English Structure Planning*, London, Pion.

Crossman, R. (1975) *Diaries of a Cabinet Minister*, London, Cape.

Crow, S., Tewdwr-Jones, M. and Harris, N. (1997) The modification phase of Local Plan preparation, *Journal of Planning and Environment Law*, April, pp. 291–313.

Cullingworth, B. (1994) Fifty years of post-war planning, *Town Planning Review*, 65, pp. 277–303.

Cullingworth, B. (1996a) Fifty years of the 1947 Act, *Town and Country Planning*, 66, pp. 130–56.

Cullingworth, B. (1996b) A vision lost, *Town and Country Planning*, 65, pp. 172–4.

Cullingworth, B. (1997) British land-use planning: a failure to cope with change? *Urban Studies*, 34, pp. 945–60.

Cullingworth, B. (1999) British planning: positive or reluctant? in *British Planning: Fifty Years of Urban and Regional Policy*, edited by B. Cullingworth, London, The Athlone Press, pp. 276–82.

Cullingworth, J. and Nadin, V. (2002) *Town and Country Planning in the UK*, London, Routledge.

Curran, J., Wood, C. and Hilton, M. (1998) Environmental appraisal of UK development plans, *Environment and Planning B*, 25, pp. 411–33.

Curry, N. and McNab, A. (1986) Development control and landscape protection, *Countryside Planning Yearbook*, 7, pp. 89–115.

Davies, H. (1998) The evolution of the British planning system 1947–97, *Town Planning Review*, 69, pp. 135–52.

Dawkins, C. and Nelson, A. (2002) Urban containment policies and housing prices: an international comparison with implications for future research, *Land Use Policy*, 19, pp. 1–12.

Deegan, J. (2004) No re-kindling here: the planning profession deserves better from draft PPS1, *Town and Country Planning*, 73, p. 118.

Delafons, J. (1995) Policy forum: planning research and the policy process, *Town Planning Review*, 66, pp. 83–109.

Department of the Environment (1975) *Review of the Development Control System*, Circular 113/75, London, HMSO.

Department of the Environment (1976) *The Dobry Report: Action by Local Authorities*, Circular 9/76, London, HMSO.

Department of the Environment (1984) *Memoranda on Structure and Local Plans*, Circular 22/84, London, HMSO.

Department of the Environment (1988) *PPG12: Structure and Local Plans*, London, HMSO.

Department of the Environment (1992) *Evaluating the Effectiveness of Planning: A Report by PIEDA and D. Diamond*, London, HMSO.

Diamond, D. (1995) Geography and Planning in the Information Age, *Transactions of the Institute of British Geographers*, new series, 20, pp. 131–8.

Dobry, G. (1975) *Review of Development Control: Final Report*, London, HMSO.

DTLR (2001) *Planning Green Paper: Planning: Delivering a Fundamental Change*, London, Stationery Office and DTLR Web site (www.dtlr.gov.uk).

Dunn W. (1994) *Public Policy Analysis*, Englewood Cliffs, NJ, Prentice Hall.

Elson, M. (1981) Structure Plan policies for pressured rural areas, *Countryside Planning Yearbook, 2*, pp. 49–70.

Elson, M., Steenberg, C. and Downing, L. (1998) *Rural Development and Land Use Planning Policies*, Salisbury, Rural Development Commission.

Elson, M., Walker, S. and McDonald, R. (1993) *The Effectiveness of Green Belts: A Report for the Department of the Environment*, London, HMSO.

Entec, Eftec and Richard Hopkinson Consultancy (2004) *Study into the Environmental Impacts of Increasing Housing Supply in the UK*, London, DEFRA.

Etzioni, A. (1993) *The Spirit of Community Rights, Responsibilities and the Communitarian Agenda*, New York, Crown Publishers.

Evans, A. (1992) Rabbit hutches on postage stamps: planning, development and political economy, *Urban Studies*, 28, pp. 853–70.

Eversley, D. (1973) *The Planner in Society*, London, Faber.

Fischer, F. (2003) *Reframing Public Policy: Discursive Politics and Deliberative Practices*, Oxford, Oxford University Press.

Flyvberg, B. (1998) *Rationality and Power: Democracy in Practice*, Chicago, University of Chicago Press.

Foucault, M. (1982) The subject and power, in *Michel Foucault: Beyond Structuralism and Hermeneutics*, edited by H. Dreyfus and P. Rabinow, Brighton, Harvester Press, pp. 208–26.

Friedmann, J. (1973) *Retracking America: A theory of Transactive Planning*, New York, Doubleday Anchor.

Gasper, D. and George, R. (1998) Analysing argumentation in planning and public policy: assessing, improving and transcending the Toulmin model, *Environment and Planning B*, 25, pp. 367–90.

Gilg, A. (1978) *Countryside Planning*, Newton Abbot, David and Charles.

Gilg, A. (1991) *Countryside Planning Policies for the 1990s*, Wallingford, CAB International.

Gilg, A. (1996) *Countryside Planning*, 2nd edn, London, Routledge.

Gilg, A. (2000) The public examination into the draft RPG for SW England, *Town and Country Planning*, 69, pp. 174–6.

Gilg, A. and Blacksell, M. (1977) Planning control in an Area of Outstanding Natural Beauty, *Social and Economic Administration*, 11, 3, pp. 206–15.

Gilg, A. and Blacksell, M. (1981) *The Countryside: Planning and Change*, London, Allen and Unwin.

Gilg, A. and Kelly, M. (1996a) Closing the agricultural loophole, *Town and Country Planning*, 65, pp. 337–9.

Gilg, A. and Kelly, M. (1996b) The analysis of development control decisions, *Town Planning Review*, 67, pp. 203–28.

Gilg, A. and Kelly, M. (1997a) Rural planning in practice: the case of agricultural dwellings, *Progress in Planning*, 47, pp. 75–157.

Gilg, A. and Kelly, M. (1997b) The delivery of planning policy in Great Britain: explaining the implementation gap, *Environment and Planning C*, 15, pp. 19–36.

Gilg, A. and Kelly, M. (2000) Managing change in the millennium, *Town Planning Review*, 71, pp. 269–88.

Greed, C. (ed.) (1996) *Implementing Town Planning: The Role of Town Planning in the Development Process*, Harlow, Longman.

Gregory, D. (1970) *Green Belts and Development Control*, Birmingham, Centre for Urban and Regional Studies, University of Aston.

Habermas. J. (1984) *The Theory of Communicative Action*, Cambridge, Polity Press.

Hague, C. (1998) Creative Instincts, *Planning*, 5 June, p. 23.

Hague, C. (2000) What is planning and what do planners do? in *Introduction to Planning Practice*, edited by P. Allmendinger, A. Prior and J. Raemaekers, Chichester, Wiley, pp. 1–20.

Hague, C. (2002) A response comment to T. Roberts, The seven lamps of planning, *Town Planning Review*, 73, pp. 1–15.

Hales, R. (2000) Land Use Development Planning and the notion of sustainable development, *Journal of Environmental Planning and Management*, 43, pp. 99–121.

Hall, C., McVitie, A. and Moran, D. (2004) What does the public want from agriculture and the countryside? A review of evidence and methods, *Journal of Rural Studies*, 20, pp. 11–25.

Hall, P. (1974) The containment of urban England, *Geographical Journal*, 140, pp. 386–418.

Hall, P. (1988) The industrial revolution in reverse? *The Planner*, 74, pp. 15–19.

Hall, P. (1992) *Urban and Regional Planning*, 3rd edn, London, Routledge.

Hall, P., Thomas, R., Gracey, H. and Drewett, R. (1973) *The Containment of Urban England*, London, Allen and Unwin.

Hardy, D. (1991) *Campaigning for Town and Country Planning I, From Garden Cities to New Towns 1899–1946; II, From New Towns to Green Politics*, London, Spon.

Hardy, D. and Ward, C. (1984) Lessons from the plotlands, *The Planner*, 70, pp. 17–20.

Haughton, G. and Counsell, D. (2002) Going through the motions? Transparency and participation in English regional planning, *Town and Country Planning*, 71, pp. 120–3.

Healey, P. (1983) *Local Plans in British Land Use Planning*, London, Pergamon Press.

Healey, P. (1988) The British planning system and managing the urban environment, *Town Planning Review*, 59, pp. 393–417.

Healey, P. (1991) Researching planning practice, *Town Planning Review*, 62, pp. 447–59.

Healey, P. (1992) The communicative work of development plans, *Environment and Planning B*, 20, pp. 83–104.

Healey, P. (1997) *Collaborative Planning: Shaping Places in Fragmented Societies*, Basingstoke, Macmillan.

Healey, P. (1998) Collaborative planning in a stakeholder society, *Town Planning Review*, 69, 1, pp. 1–21.

Healey, P. (1999) Deconstructing communicative planning theory: a reply to Allmendinger and Tewdwr-Jones, *Environment and Planning A*, 31, pp. 1129–35.

Healey, P. and Shaw, T. (1994) Changing meanings of 'environment' in the British planning system, *Transactions of the Institute of British Geographers*, new series, 19, pp. 425–38.

Healey, P., McNamara, P., Elson, M. and Doak, A. (1988) *Land Use Planning and the Mediation of Urban Change*, Cambridge, Cambridge University Press.

Herrington, J. (1984) Green Belt red herring, *Town and Country Planning*, 53, pp. 68–70.

Hobbs, P. (1992) The economic determinants of post-war British town planning, *Progress in Planning*, 38, pp. 179–300.

Holstein, T. (2002) Moving to a SEA Green Paper, *Town and Country Planning*, September, pp. 218–20.

Housing and Local Government, Ministry of, Planning Advisory Group (1965) *The Future of Development Plans*, London, HMSO.

Housing and Local Government, Ministry of (1969) *Committee on Participation in Planning: People and Planning*, chairman A.M. Skeffington, London, HMSO.

Housing and Local Government, Ministry of (1970) *Development Plans: A Manual on Form and Content*, London, HMSO.

Howlett, M. and Ramesh, M. (1995) *Studying Public Policy: Policy Cycles and Policy Subsystems*, Toronto, Oxford University Press.

Hull, A. and Vigar, G. (1998) The changing role of the Development Plan in managing change, *Environment and Planning C*, 16, pp. 379–94.

Jarvis, R. (1996) Structure planning policy and strategic planning guidance in Wales, in *British Planning Policy in Transition: Planning in the 1990s*, edited by M. Tewdwr-Jones, London, UCL Press, pp. 43–60.

Jones, C. and Bull, T. (1997) Analysis of changing trends in UK Environmental Statements, *Journal of Planning and Environment Law*, December, pp. 1091–103.

Jones, R., Goodwin, M., Jones, M. and Simpson, G. (2004) Devolution, state personnel and the production of new territories of governance in the United Kingdom, *Environment and Planning A*, 36, pp. 89–109.

Jordan, A. (ed.) (2002) *Environmental Policy in the European Union*, London, Earthscan.

Kelly, M., Selman, P. and Gilg, A. (2004) Taking sustainability forward: relating practice to policy in a changing legislative environment, *Town Planning Review*, 75, pp. 309–35.

Keyes, J. (1986) Controlling residential development in the Green Belt: a case study, *The Planner*, 72, pp. 18–20.

Kitchen, T. (1996) A future for strategic planning policy, in *British Planning Policy in Transition: Planning in the 1990s*, edited by M. Tewdwr-Jones, London, UCL Press, pp. 124–36.

Kitchen, T. (1997) *People, Politics, Policies and Plans: The City Planning Process in Contemporary Britain*, London, Chapman.

Kitchen, T. (1998) Time to rethink the system, *Town and Country Planning*, 67, pp. 98–9.

Larkham, P. (1992) Conservation and the changing urban landscape, *Progress in Planning*, 37, pp. 84–181.

Larkham, P. and Jones, A. (1993) The character of conservation areas in Britain, *Town Planning Review*, 64, pp. 395–413.

Lichfield, N. (1996) *Community Impact Evaluation*, London, UCL Press.

Lindblom, C. (1959) The science of muddling through, *Public Administration Review*, 99, 79–99.

Lock, D. (2002) Planning reform-action after the Green Paper, *Town and Country Planning*, 71, pp. 252–4.

Lowe, P. and Murdoch, J. (2003) Mediating the 'national' and the 'local' in the environmental policy process: A case study of the CPRE, *Environment and Planning C*, 21, pp. 761–78.

MacGregor, B. and Ross, A. (1995) Master or servant? The changing role of the development plan in the British planning system, *Town Planning Review*, 66, pp. 41–59.

Madanipour, A., Hull, A. and Healey, P. (eds) (2001) *The Governance of Place: Space and Planning Processes*, Aldershot, Ashgate.

Mantysalo, R. (2002) Dilemmas in critical planning theory, *Town Planning Review*, 73, pp. 417–36.

Marshall, T. (2002) The retiming of English regional planning, *Town Planning Review*, 73, pp. 171–95.

McAuslan, P. (1980) *The Ideologies of Planning Law*, Oxford, Pergamon Press.

McCarthy, P. for Prism Research (1995) *Attitudes to Town and Country Planning*, London, HMSO.

McDonald, G. (1989) Rural land use decisions by bargaining, *Journal of Rural Studies*, 2, pp. 325–35.

McLoughlin, J. (1973) *Control and Urban Planning*, London, Faber.

McNamara, P. and Healey, P. (1984) The limitations of development control data in planning research, *Town Planning Review*, 55, pp. 91–7.

Meikle, J., Pattinson, M. and Zetter, J. (1991) Breakdown of housing development costs, *The Planner*, 16 April, p. 7.

Midgeley, J. (2000) Exploring alternative methodologies to establish the effects of land area designations in development control decisions, *Planning Practice and Research*, 15, pp. 319–33.

Monk, S. and Whitehead, C. (1999) Evaluating the economic impact of planning controls in the UK, *Land Economics*, 75, pp. 74–93.

Morrison, N. and Pearce, B. (2000) Developing indicators for evaluating the effectiveness of the UK land use planning system, *Town Planning Review*, 71, pp. 191–211.

Moseley, M. (2003) *Rural Development: Principles and Practice*, London, Sage.

Moss, G. (1978) The village: a matter of life and death, *Architects Journal*, 167, pp. 100–39.

Munton, R. (1983) *London's Green Belt: Containment in Practice*, London, Allen and Unwin.

Murdoch, J. (2000) Space against time: competing rationalities in planning for housing, *Transactions of the Institute of British Geographers*, new series, 25, pp. 503–19.

Murdoch, J. (2004) Putting discourse in its place: planning sustainability and the urban capacity study, *Area*, 36, pp. 50–8.

Murdoch, J. and Abram, S. (2002) *Rationalities of Planning*, Aldershot, Ashgate.

Murdoch, J. and Lowe, P. (2003) The preservationist paradox: modernism, environmentalism and the politics of spatial division, *Transactions of the Institute of British Geographers*, new series, 28, pp. 318–32.

Murdoch, J. and Marsden, T. (1994) *Reconstituting Rurality*, London, UCL Press.

Murdoch, J. and Tewdwr-Jones, M. (1999) Planning and the English regions, *Environment and Planning C*, 16, pp. 715–29.

Murdoch, J., Abram, S. and Marsden, T. (1999) Modalities of planning: a reflection on the persuasive powers of the Development Plan, *Town Planning Review*, 70, pp. 191–212.

Murdoch, J., Lowe, P., Ward, N. and Marsden, T. (2003) *The Differentiated Countryside*, London, Routledge.

Norcliffe, G. and Hoare, A. (1982) Enterprise Zones for the inner city, *Area*, 14, pp. 265–74.

Office of National Statistics (2002) *National Population Projections: 2000 Based*, London, Stationery Office.

Office of the Deputy Prime Minister (2004) *Land Use Change in England: Residential Development to 2003*, London, The Office.

O'Riordan, T. (1985) Future directions for environment policy, *Environment and Planning A*, 17, pp. 1431–46.

O'Riordan, T. (1995) Environmental science on the move, in *Environmental Science for Environmental Management*, edited by T. O'Riordan, Oxford, Pergamon Press, pp. 1–16.

Pacione, M. (1990) Private profit and public interest in the residential development process: a case study of conflict in the urban fringe, *Journal of Rural Studies*, 6, pp. 103–16.

Pacione, M. (1991) Development pressure and the production of the built environment in the urban fringe, *Scottish Geographical Magazine*, 107, pp. 162–9.

Pearce, M. (1992) The effectiveness of the British land use planning system, *Town Planning Review*, 63, pp. 13–28.

Pecol, E. (1996) GIS as a tool for assessing the influence of countryside designations on landscape change, *Journal of Environmental Management*, 47, pp. 355–67.

Pennington, M. (2000) Public choice theory and the politics of urban containment: voter-centred versus special-interest explanations, *Environment and Planning C*, 18, pp. 145–62.

Pennington, M. (2002) *Liberating the Land: The Case for Private Land Use Planning*, London, Institute of Economic Affairs.

Pickering, M. (1987) Development control topics, *The Planner*, 73, pp. 94–8.

Pountney, M. and Kingsbury, P. (1983) Aspects of development control. The relationship with local plans, *Town Planning Review*, 54, pp. 139–54.

Poxon, J. (2001) Shaping the planning profession of the future: the role of planning education, *Environment and Planning B*, 28, pp. 563–80.

Preece, R. (1981) *Problems of Development Control in the Cotswolds AONB*, Oxford, School of Geography, Oxford University.

Preece, R. (1990) Development control studies: scientific method and policy analysis, *Town Planning Review*, 61, pp. 59–74.

Pressman, J. and Wildavsky, A. (1973) *Implementation*, Berkeley, CA, University of California Press.

Punter, J. and Bell, A. (1998) The appeal of design, *Town and Country Planning*, 67, pp. 179–81.

Punter, J. and Bell, A. (1999) The role of policy in design appeals, *Town Planning Review*, 69, pp. 231–58.

Purton, P. and Douglas, C. (1982) Enterprise zones in the UK, *Journal of Planning and Environment Law*, April, pp. 412–22.

Putnam, R., Leonardi, R. and Nanetti, R. (1993) *Making Democracy Work: Civic Traditions in Modern Italy*, Princeton, NJ, Princeton University Press.

Quality Assurance Agency for Higher Education (2002) *Town and Country Planning Subject Benchmark Statement*, Gloucester, Quality Assurance Agency.

Rawls, J. (1993) *Political Liberalism*, New York, Columbia University Press.

Richardson, T. (2001) The pendulum swings again: in search of new transport rationalities, *Town Planning review*, 72, pp. 299–319.

Roberts, P. (1996) Regional planning guidance in England and Wales: back to the future? *Town Planning Review*, 67, 1, pp. 97–109.

Roberts, T. (2002) The seven lamps of planning, *Town Planning Review*, 73, 1, pp. 1–15.

Rowswell, A. (1989) Rural depopulation and counterurbanisation: a paradox, *Area*, 21, pp. 93–4.

Royal Commission on Environmental Pollution (2002) *Environmental Planning*, London, Stationery Office.

Rydin, Y. (1985) Residential development and the planning system: a study of the housing land system at the local level, *Progress in Planning*, 24, pp. 3–69.

Rydin, Y. (1993) *The British Planning System: An Introduction*, Basingstoke, Macmillan.

Rydin, Y. (1998) *Urban and Environmental Planning in the UK*, Basingstoke, Macmillan.

Rydin, Y. (2004) So what does PPS1 tell us about spatial planning? *Town and Country Planning*, 73, p. 117.

Sager, T. (2001a) Planning style and agency properties, *Environment and Planning A*, 33, pp. 509–32.

Sager, T. (2001b) Positive theory of planning: the social choice approach, *Environment and Planning A*, 33, pp. 629–47.

Sager, T. (2001c) Manipulative features of planning styles, *Environment and Planning A*, 33, pp. 765–81.

Schrader, H. (1994) Impact assessment of the EU structural funds to support regional economic development in rural areas of Germany, *Journal of Rural Studies*, 10, pp. 357–65.

Scott, A. (2001) Special Landscape Areas: their operation and effectiveness in Ceredigion, Wales, *Town Planning Review*, 72, pp. 469–80.

Scott report (1942) *Report of the Committee on Land Utilisation in Rural Areas*, Cmd 6378, London, HMSO.

Selman, P. (1988) Rural land-use planning: resolving the British paradox, *Journal of Rural Studies*, 4, pp. 277–94.

Short, J., Fleming, S. and Witt, S. (1986) *Housebuilding, Planning and Community Action*, London, Routledge.

Short, J., Witt, S. and Fleming, S. (1987) Conflict and compromise in the built environment: housebuilding in central Berkshire, *Transactions of the Institute of British Geographers*, new series, 12, pp. 29–42.

Simmons, M. (1999) The revival of regional planning, *Town Planning Review*, 70, pp. 159–72.

Sinclair, G. (1992) *The Lost Land: Land Use Change in England 1945–1990*, London, Council for the Protection of Rural England.

Spencer, D. (1997) Counterurbanisation and rural depopulation revisited: landowners, planners and the rural development process, *Journal of Rural Studies*, 13, pp. 75–92.

Stead, D. (2001) Relationships between land use, socio-economic factors, and travel patterns in Britain, *Environment and Planning B*, 28, pp. 499–528.

Sutcliffe, A. (1981) *British Town Planning: The Formative Years*, Leicester, Leicester University Press.

Tait, M. and Campbell, H. (2000) The politics of communication between planning officers and politicians, *Environment and Planning A*, 32, pp. 489–506.

Taylor, N. (1998) *Urban Planning Theory since 1945*, London, Sage.

Tewdwr-Jones, M. (1994) The Development Plan in policy implementation, *Environment and Planning C*, 12, 2, pp. 145–63.

Tewdwr-Jones, M. (1995) Development control and the legitimacy of planning decisions, *Town Planning Review*, 66, pp. 163–81.

Tewdwr-Jones, M. (1996) Land-use planning policy after Thatcher, in *British Planning Policy in Transition: Planning in the 1990s*, edited by M. Tewdwr-Jones, London, UCL Press, pp. 1–14.

Tewdwr-Jones, M. (1998) Planning modernised? *Journal of Planning and Environment Law*, June, pp. 519–28.

Tewdwr-Jones, M. (1999) Discretion, flexibility and certainty in British planning: emerging ideological conflicts, *Journal of Planning Education and Research*, 18, pp. 244–56.

Tewdwr-Jones, M. (2002) *The Planning Polity: Planning Government and the Policy Process*, London, Routledge.

Tewdwr-Jones, M. and Allmendinger, P. (1998) Deconstructing communicative rationality: a critique of collaborative planning, *Environment and Planning A*, 30, pp. 1975–89.

Tewdwr-Jones, M. and Thomas, H. (1998) Collaborative action in local plan making: planners' perceptions of planning through debate, *Environment and Planning B*, 25, pp. 127–44.

Therivel, J., Doak, J. and Stott, M. (1998) Desperately seeking sustainability, *Town and Country Planning*, 67, pp. 26–8.

Thomas, H. (1996) Public participation in planning, in *British Planning Policy in Transition: Planning in the 1990s*, edited by M. Tewdwr-Jones, London, UCL Press, pp. 168–88.

Thomas, K. (1997) *Development Control: Principles and Practice*, London, UCL Press.

Thomas, K. and Roberts, P. (2000) Metropolitan strategic planning in England, *Town Planning Review*, 71, pp. 25–49.

Thornley, A. (1993) *Urban Planning under Thatcherism: The Challenge of the Market*, London, Routledge.

Toulmin, S. (1958) *The Uses of Argument*, Cambridge, Cambridge University Press.

Town and Country Planning Association (1999) *Your Place and Mine: Reinventing Planning*, London, The Association.

Underwood, J. (1982) Local plans: the state of play, *The Planner*, 68, pp. 136–53.

Urban Task Force (1999) *Towards an Urban Renaissance, Lord Rogers*, London, Department of the Environment, Transport and the Regions.

Uthwatt report (1942) *Final Report of the Expert Committee on Compensation and Betterment*, Cmd 6386, London, HMSO.

Vigar, G. and Healey, P. (2002) Developing environmentally respectful policy programmes: five key principles, *Journal of Environmental Planning and Management*, 45, pp. 517–32.

Vigar, G., Healey, P., Hull, A. and Davoudi, S. (2000) *Planning Governance and Spatial Strategy in Britain: An Institutionalist Approach*, Basingstoke, Macmillan.

Ward, S. (2004) *Planning and Urban Change*, 2nd edn, London, Sage.

Watkins, C., Wale, C., Haines-Young, R. and Murdock, A. (2001) Contextualising development pressure: the use of GIS to analyse planning applications in the Sussex Downs Area of Outstanding Natural Beauty, *Town Planning Review*, 72, pp. 373–91.

Webster, C. (1998) Analytical public choice planning, *Town Planning Review*, 69, pp. 191–209.

Whatmore, S. and Boucher, S. (1993) Bargaining with nature: the discourse and practice of 'environmental planning gain', *Transactions of the Institute of British Geographers*, new series, 18, pp. 166–78.

While, A., Jonas, A. and Gibbs, D. (2004) Unblocking the city? Growth pressures, collective provision, and the search for new spaces of governance in greater Cambridge, England, *Environment and Planning A*, 36, pp. 279–304.

Whitehand, J. and Carr, C. (1999) The changing fabric of ordinary residential areas, *Urban Studies*, 36, pp. 1661–78.

Whitehand, J. and Morton, N. (2003) Fringe belts and the recycling of urban land: an academic concept and planning practice, *Environment and Planning B*, 30, pp. 819–39.

Willis, K. and Powe, N. (1995) Planning decisions on waste disposal sites, *Environment and Planning B*, 22, pp. 93–107.

Wood, R. (2000) Using appeal data to characterise local planning authorities, *Town Planning Review*, 71, pp. 97–107.

Wood, C., Dipper, B. and Jones, C. (2000) Auditing the Assessment of the Environmental Impacts of Planning Projects, *Journal of Environmental Planning and Management*, 43, pp. 23–47.

Wood, R., Handley, J. and Bell, P. (1998) An analysis of Planning Inspectorate records, *Journal of Planning and Environment Law*, November, pp. 1007–27.

Woods, M. (1997) Advocating rurality? The repositioning of rural local government, *Journal of Rural Studies*, 14, pp. 13–26.

Index

Abram, S. 17, 93
Action Area Plans 15, 18–20, 90, 193
Actor-networks 89, 97–102, 129, 173 *see also* key actors; networks
Adams, D. 90–1, 116
Advocacy planning 65
Affordable (social) housing 39, 148, 150
Agricultural census 112–14
Agricultural exemptions and exceptions from planning control 43, 106
Agriculture Act 1947, 9
Alder, J. 57
Alexander, E. 64
Allmendinger, P. 27, 160–1, 166, 172
Amenity societies 144
Anderson, M. 38, 40, 141–2, 143
AONB *see* Areas of Outstanding Natural Beauty
Appeal-led planning era 102, 163
Appeals 40–2, 102–5, 121
Application quality 131
Arbitration 60
Area-wide Local Plans 197
Areas of Outstanding Natural Beauty (AONB) 42, 43, 136–42, 143, 193
Article 4 Direction Orders 44
Audit Commission 104, 155

Baker, M. 24, 90
Bargaining 38, 39, 60, 64, 65
Barker, K. 148–50
Barlow Report 8
Barrass, R. 89
Bartlett, W. 197
Bedford, T. 99
Behaviouralism 72, 106
Benchmarks 66
Best, R. 112–114
Best practicable environmental option 60
Best value 73, 94, 185

Betterment and Compensation 8–9, 156
Bibby, P. 115–16, 152, 186, 187
Bingham, M. 104
Bishop, K. 86
Blacksell, M. 112, 135–42
Blair, Tony 53, 73, 160, 161, 162, 194
Blight Notice 46
Blowers, A. 168–9
Booth, P. 46, 47, 109, 197
Boucher, S. 168
Bracken, I. 89
Bramley, G. 129, 146, 197
Breheny, M. 61, 119
Brindley, T. 164
Bristow, M. 151–2
British democratic system 52
Brotherton, I. 104, 131, 133, 134
Brownfield land 93, 114, 115, 116, 122, 129, 142–4, 150
Bruff, G. 90
Bruton, M. 90, 91
Building Preservation Notice 44
Bull, T. 37
Buller, H. 100
Bunnell, G. 106
Burton, E. 119
Business Planning Zones 194

Café culture 184, 188, 196
Call-in 18, 42
Cambridge Architectural Research 124
Campbell, H. 39, 106–7, 108, 145
Campbell, S. 56
Canary Wharf 44
Capital accumulation 99
Cardiff Bay 44
Carmona, M. 1, 66, 73, 74, 109, 189
Carr, C. 142–4
Carter, J. 90
Cause and effect 67–72
Certificate of Lawfulness of Existing Use or Development 46

Certficicate of Lawfulness of Proposed Use or Development 46
Champion, T. 113, 125–7
Checklists 36–7
Cherry, G. 47, 95, 163
Cheshire, P. 152
Christie, I. 188
Circulars from Central Government
Department of the Environment (DDE) 71/74, 112
22/80, 35, 61, 109
14/84, 120
22/84, 14, 23
16/87, 35
16/91, 38
11/95, 38
Ministry of Housing and Local Government 42/55, 120
Claydon, J. 19, 34, 109
Cloke, P. 68, 106, 109, 150–2
Coalition building 177
Coastal Preservation Area 103
Coates, D. 18
Cognitive continuum 107
Coleman, A. 112
Collaborative Planning 64, 162, 170, 171
Command Planning 64
Common Agricultural Policy 118
Communicative planning (theory) 64, 66, 106, 170
Communists 59
Communitarianism 54
Community Strategies 26, 192, 193
Involvement, Statement of 193
Comprehensive linear systems planning model 62–4
Compulsory Purchase 9
Conceptual evaluations 79, 166–9
Conditional Permission 38–9
Conflict management 171, 172
resolution 55–7, 59–60, 177, 181–2
Conservation Areas 43, 44, 142–5, 193
culture 59

Conservative (Tory) Party 55, 102, 121
Consortium Developments 120–1
Constraints on planning 68–9
Conventions theory 101
Corkindale, J. 68
Council for the Protection of Rural England (CPRE) 7, 93, 114, 129, 150
Council Housing 6, 10, 11, 115
Councillors 39, 106–7
Counsell, D. 28, 87, 90
Countryside Commission 188
Countryside types: preserved; contested; paternalist; client list 101–2
County Councils 11–14, 88–9, 125
Covenants 61
CPRE see Council for the Protection of Rural England
Critical planning theory 170
Crook, A. 148
Cross, D. 151–2
Crossman, R. 42
Crow, S. 21, 122, 123
Cullingworth, B. 25, 26, 155–6, 185
Culture, Media and Sport, Department for 86
Curran, J. 90
Curry, N. 134, 135, 152

Davies, H. 156
Davoudi, S. 109
Dawkins, C. 148
Decentralisation 116–20
Deconstruction 72, 92
Deegan, J. 195
DEFRA see Environment and Rural Affairs, Department of
Delafons, J. 55
Delivery systems 62–5
Deputy Prime Minister, Office of the 58,115
Derrida, J. 72
Designated landscapes 133–42, 178
Development control 12, 14, 18, 32–47, 96–109, 130–45, 177, 178, 183, 190, 192, 194, 197
definition of 32–33
Development Plans 9–31, 42, 95, 175–6, 179
Diamond, D. 76, 155
Dipper, B. 109
District Councils 11–14, 88–9
District Plans 15, 18–20, 90

Doak, A. 83
Dobry Report 108–9, 177
Douglas, C. 43
Dunn, W. 74–5

e-retailing 188
e-working 188
Ecological Modernisation 168–9
Ecologists 59
Economic cycles 159–60
EIP see Examination In Public
Elites 52–3, 54
Elson, M. 83, 89, 91, 120, 133, 151
Empiricism 70
Employment 118, 126
Enforcement Controls 45
Ennis, F. 109
Entec 150
Enterprise Zones 43–4
Environment, Department of 76, 77, 152
Environment and Rural Affairs, Department of 86
Environmental Appraisal 24, 90
Environmental (Impact) Assessment 37
Environmental determinism 165
Etzioni, A. 54
European (EU) agenda 194
European legislation 37
European Spatial Development Perspective 27
Evans, A. 117
Examination in Public (EIP) 14–17, 89, 93
Externalities 60

Farm census 112–14
Fischer, F. 72, 170
Flyvberg, B. 92
Foley, P. 152
Forecasting the future 89, 123, 150, 183–8
Foucault, M. 51, 72, 92, 93, 107, 125, 170, 172
Friedmann, J. 64
Friends of the Earth 129

Gallent, N. 109
Gasper, D. 73–5
GDO see General Permitted Development Order
General Information Systems (GIS) 132
General Permitted Development Order (GDO) 32–3, 43
George, R. 73–5
Getting things done 91, 94, 166

Giddens, A. 70, 170
Gilg, A. 13, 28, 39, 43, 57, 62, 63, 67, 69, 83, 106, 112, 129, 130–1, 135–42, 152, 189
Gilg-Selman spectrum 62, 157
GIS see General Information Systems
Glasson, J. 109
Golf 41, 101
Governance 93, 168, 172
Governmentality 93
Green Belts 42, 93, 102, 115, 120–1, 123, 124, 127, 132–3, 141, 146, 157, 180
Green Paper on Planning 191–2 (2001)
Gregory, D. 131
Growth management 168, 170

Habermas, J. 51, 92, 106, 170, 172
Hague, C. 72, 159
Hales, R. 90
Hall, P. 1, 112, 116–20, 145
Hardy, D. 7
Haughton, G. 87
Healey, P. 1, 21, 83, 91, 92, 109, 132, 134, 157, 160, 169, 170, 171, 172
Heath, Edward 53
Heathrow Airport 56
Herrington, J. 121
Historical evaluations 79, 157–63
Hoare, A. 43
Hobbs, P. 159–60
Hoggart, K. 100
Holstein, T. 90
House builders 98, 116, 120–1
House Builders Federation 14, 93
House building 11, 115, 145–50, 175, 181
House prices 145–50, 179, 180, 182, 195
Households 187
Housing, density of 115
Housing and Local Government, Ministry of 11, 14
Housing, Town Planning Act 1909, 6
1919, 6
Howlett, M. 49, 83
Hull, A. 89, 109
Hume, D. 89
Hutchinson, J. 152
Hybrid approach 79–82, 133, 173–84

Ideal planning system 63
Immigration 188

Implementation gap 67–72, 108, 152
Incremental planning 65
Individual rights 46
Inferential assessments 79, 154–7
Innes, J. 170
Interest rates 146

Jarvis, R. 30, 188
Jones, A. 44
Jones, C. 37, 109
Jones, R. 87
Jordan, A. 163–4

Kelly, M. 39, 43, 67, 106, 130–1, 152, 189, 194, 195
Key actors 131 see also actor networks
Key villages/settlement policies 106, 136–42, 152
Keyes, J. 121
Kingsbury 132, 133
Kitchen, T. 94, 95, 107
Klosterman, R. 83
Knowledge 69–72

Labour Party 55, 87, 122
Lambert, C. 197
Land ownership 56–57, 61, 91
Land prices/values 117, 145–50, 155, 156, 157
Land use change 112–20, 121
statistics 115–16, 121, 141
Land uses, separation of 117, 155, 178, 179, 181, 185
Landscape restraint/designation policies 100–1, 134
Larkham, P. 44, 144–5
LDFs see Local Development Frameworks
Leisure Plots 44
Leverage planning 164
Liberal Democratic Party 55, 122
Lichfield, N. 71, 76
Lindblom, C. 67
Listed Buildings 44–5, 144–5
Little, J. 109
Littlewood, S. 109
Local Development Documents 194
Frameworks (LDFs) 192, 193, 194
Schemes 194
Local Government, Planning and Land Act 1980, 18, 43
Local Government Reorganisation 11–14, 21, 24–5
Local Plans 15, 18–22, 24, 31, 90–2, 103, 192, 194, 197

Logical positivists/positivism 70
Lowe, P. 93, 128

MacGregor, B. 91
McAuslan, P. 56
McCarthy, P. 156
McDonald, G. 64
McLoughlin, J. 92
McNab, A. 134, 135
McNamara, P. 83, 132, 134
Madanipour, A. 172
Major, John 53, 160, 162
Making places 171
Making things happen see management of change
Management of change (Getting things done) 91, 94, 166, 173–4, 177, 180
Mantysalo, R. 170
Manual on Development Plans 14–15
Market failure 57, 60, 61, 166
forces 61
Market-critical planning 164
Market-led planning 164
Marsden, T. 100, 132
Marshall, T. 88, 107, 108
Material considerations 36, 109, 190
May, H. 90–1
Megalopolis 114, 117–18
Meikle, J. 145
Metropolitan Areas/Authorities 21, 27, 88
Midgeley, J. 134, 136
Minerals Plans 24
Middle class (self-interest) 100–1, 151, 173
Mixed economy 59–60
Monk, S. 146–8, 149
Morrison, N. 49, 76, 77, 83, 164
Mortgage rates 146
Morton, N. 143–4
Moseley, M. 76
Moss, G. 127, 129
Munton, R. 120
Murdoch. J. 17, 51, 86, 92–3, 100–1, 125, 128, 131, 151

Nadin, V. 25, 26, 185
National Land Use Database 116
National Park Plans 24, 31
National Parks 31, 42, 43, 134, 138–9, 193
National Parks and Access to the Countryside Act 1949, 9
National Statistics, Office of 128, 188

Negative planning 10
Neighbourhood interests 176
and village plans 193
Nelson, A. 148
Networks see also actor-networks 129
New Labour 87, 121, 160, 161, 166
New Left 165
New Right 107, 160, 161, 165, 166
New towns 184
New Towns Act 1946, 9
Nicholson, D. 90, 91
NIMBY (Not in my backyard) 100–1, 103, 118, 122, 124
Non-statutory plans 90
Norcliffe, G. 43

Obligations 38, 39
Oil Refinery 42
Ordnance Survey data on land use 112, 114, 115
O'Riordan, T. 59, 60, 69, 70
Outline planning application 33

Pacione, M. 99
Participant planning 6
Partnership schemes/planning 20, 89, 164
Paternalism 57
Peacehaven 7
Pearce, B. 49, 76, 77, 83, 164
Pearce, M. 153, 155
Pecol, E. 141
Pennington, M. 61, 119
Performance targets (criteria) 49, 73, 171
Pickering, M. 153, 155
PIEDA consultants 76
Place making 171, 172
Plan-led (planning) system 24, 36, 91, 61, 162, 163, 166, 170, 197
Plan-making 9–31, 85–96, 109, 176, 177, 178, 181, 183, 190, 191, 192
Plan, manage and monitor 160
Planning
application see Development control
balancing role of 55–9, 80–1, 175
case law 35–6
lore 35–6
future of 188–95
objectives 55–8, 175–6, 181–2
officers 106–8, 174, 189

Planning – *continued*
  organisations involved with
    86–9
  permission *see* Development
    control
  determination of need for
    46
  refusal of 39–40, 102, 133
  as social action and practice
    72–3, 172
  social impact of 157
  and society 50–5, 173–4
  and stakeholders 65
Planning Advisory Group 11, 12
Planning Balance Sheet 76
Planning blight 46
Planning and Compensation Act
  1991, 24, 36, 38
Planning and Compulsory
  Purchase Act 2004, 194
Planning gain 39, 60, 99, 105
Planning Inspectorate 40, 102–3,
  104
Planning (Listed Buildings and
  Conservation Areas) Act,
  1990 44
Planning Policy Guidance Notes
  (PPGs) 23–4, 35
  PPG1, 194
  PPG2, 120
  PPG3, 35, 93, 129
  PPG7, 35
  PPG13, 91
Planning Policy Statements (PPS)
  194
PPS1 194, 195
Policies planning 64
Political economy 70, 99, 165, 166
  history 53
  modernisation 168–9
Politicians 51
Polluter pays 60
Positive planning 10, 94
Popular planning 164
Population change 125–30, 185,
  186
Postmodernism 72, 131
Pountney, M. 132, 133
Powe, N. 107
Power 50–5
  relations 160
Poxon, J. 189
PPG *see* Planning Policy Guidance
  Notes
Precautionary principle 59
Predict and provide planning 61,
  129, 160, 174, 182–3
Preece, R. 153–4

Pressman, J. 67
Private housing 11, 146–8
Proactive planning 60, 177
Procedural Planning Theory 64,
  165
Property rights 61
Protected landscapes 133–42, 178
Public
  interest 36, 54, 169, 182, 192
  inquiries 40–2
  participation/involvement 10,
    14, 16, 54, 60, 89, 92, 99,
    156, 176, 193, 195
  policy 50–5
Public Health Act 1848, 5
Public Local Inquiry 20
Purchase Notice 46
Purton, P. 43
Putnam, R. 54

Qualitative assessments 79, 154–7
Quality Assurance Agency for
  Higher Education 66

Rabbit hutches on postage stamps
  117
Ramesh, M. 49, 83
Rational planning model 62–4
Rawls, J. 54
Reactive planning 60, 177
Redcliffe Maud Report 11, 12
Regional Development Agencies
  (RDAs) 27, 87–8, 192
Regional economic planning 27,
  86–7
Regional planning 27–30, 86–8,
  94, 122–5, 157, 163, 174,
  178, 194
Regional Planning Assemblies 29,
  87–8, 192, 197
Regional Planning Guidance
  (RPG) 28, 87–8, 89–90,
  122–5, 162, 192
Regional Spatial Strategies (RSSs)
  192, 193, 194
Regulative planning 164
Responsive planning 164
Richardson, T. 72
Roberts, P. 22, 28
Roberts, T. 75, 162, 163
Ross, A. 91
Rowswell, A. 106
Royal Commission on
  Environmental Pollution
  (2002) 23, 27, 29, 31, 57, 74,
  189, 190, 191
Royal Town Planning Institute 7
Rural settlement 135–142

Rurban fringe 112
Rydin, Y. 66, 83, 89, 99–100, 154,
  165, 166, 167, 188, 195

Sager, T. 64, 65
Schrader, H. 68
Scott, A. 134–5
Scott Report (1942) 8
Secretary of State for the
  Environment 15
SERPLAN 122, 123
Scotland 26, 27, 29, 31
SEA *see* Strategic Environmental
  Assessment
Sellgren, J. 110
Selman, P. 62, 197
Sequential approach 130, 161
Shaw, D. 150–152
Shaw, T. 168, 169
Shepherd, J. 115–16, 152, 186,
  187
Short, J. 98–9, 103
Sieh, L. 1
Simmons, M 87
Simplified Planning Zones 43
Sinclair, G. 114–15
Skeffington Report (1969) 11,
  12
Smith, B. 109
Social housing 150
South-east England 121–5
Spatial planning (strategies) 162,
  171, 172, 190, 195
Stead, D. 119
Stone cladding 44
Strategic Environmental
  Assessment (SEA) 90, 193
Structuration, theory of 70
Structure and agency 7, 97, 116
Structure Plans 13–18, 24, 31,
  89–90, 92, 120, 150–2, 192,
  194
Subject Plans 15, 18–20, 90
Suburbs 6, 142–5, 156, 180
Sustainable Communities
  plan/programme 124, 175
Sustainable development/growth
  28, 80–1, 90, 168, 169, 175,
  177, 178, 179, 180, 184, 191,
  194, 195, 197
Sustainable Development Strategy
  (1999), UK 191
Sustainability Appraisal 193
Synoptic planning 65
Systems theory 70, 165

Tait, M. 106–7
Taylor, N. 54, 172–3, 197

Tewdwr-Jones, M. 24, 27, 47, 86, 92, 103, 109, 160–3, 166, 172
Thatcher, Margaret 52, 53, 61, 73, 86, 121, 160, 162, 163, 164
Theoretical Evaluations 166–173
Therivel, J. 109
Third-party appeals 177
Thomas, H. 16, 92
Thomas, K. 22, 54, 57, 61, 83, 94, 109, 156
Thornley, A. 160
Tory (Conservative) Party 121
Toulmin, S. 73–5
Town and Country Planning Act
  1932, 7
  1944, 9
  1947, 8, 10, 85, 101, 156
  1968, 13, 85
  1971, 13
  1990, 26
Town and Country Planning Association 6, 7, 122, 188, 189
Trade and Industry, Department of 86
Trade-offs 60

Transport issues 119–20
Transport, Local Government and the Regions, Department of 58, 86
Tree Preservation Orders 45
Trend planning 164
Typological evaluations 79, 163–166

Underwood, J. 91
Uneven regional growth 125–30
Unitary City Regions 197
Unitary Development Plans 21–2, 90, 178, 192
Urban containment 116–20, 138, 156, 178, 181
Urban Development Corporations 43
Urban Task Force 129, 188
Use Class Order (1987) 33
Uthwatt Report (1942) 8–9
Utilitarianism 54, 57, 64, 108, 168, 169
Utopianism 57

Value systems 54, 174
Vigar, G. 89, 172

Vodafone 122

Wales 26–7, 29–30, 31
Walters, Tudor 6
Ward, C. 7
Ward, S. 1, 47, 156–7, 197
Waste Plans 24
Watkins, C. 132
Webster, C. 60
Weighting planning applications 131
Welfare economics 60, 168
Welfare State 8, 9
Whatmore, S. 168
While, A. 124–5
White, A. 109
White Paper on planning (2002) 192–3
Whitehand, J. 142–4
Whitehead, C. 146–8, 149
Whitney, D. 109
Wildavsky, A. 67
Willis, K. 107
Wilson, E. 53
Wood, R. 37, 90, 103–4, 109
Woods, C. 101

Zoning 7, 60, 119